GENE
THERAPY

GENE THERAPY

Other books in the At Issue series:

GENE THERAPY

Roman Espejo, *Book Editor*

Bruce Glassman, *Vice President*
Bonnie Szumski, *Publisher*
Helen Cothran, *Managing Editor*

THOMSON
─────★───── ™
GALE

San Diego • Detroit • New York • San Francisco • Cleveland
New Haven, Conn. • Waterville, Maine • London • Munich

For more information, contact
Greenhaven Press
27500 Drake Rd.
Farmington Hills, MI 48331-3535
Or you can visit our Internet site at http://www.gale.com

LIBRARY OF CONGRESS CATALOGING-IN-PUBLICATION DATA

Gene therapy / Roman Espejo, book editor.
   p. cm. — (At issue)
Includes bibliographical references and index.
ISBN 0-7377-2256-8 (lib. : alk. paper) — ISBN 0-7377-2257-6 (pbk. : alk. paper)
   1. Gene therapy—Popular works. I. Espejo, Roman, 1977– . II. At issue (San Diego, Calif.)
RB155.8.G46173 2004
615.8'95—dc22
                                              2004040587

Contents

Introduction

Gene therapy is based on the concept that genetic disorders and acquired diseases can be treated by replacing abnormal or absent genes or by modifying their functions. Inherited disorders such as cystic fibrosis and hemophilia, as well as catastrophic diseases such as cancer and AIDS, are prospective candidates for gene therapy. Although cures for these ailments would be welcome, some medical researchers suggest that the range of diseases that can be treated with gene therapy may be limited. According to research scientist Lynn Elwell, "Only a handful of the many diseases that have a genetic basis are amenable to treatment via gene therapy: Genetic disorders caused by single genes." She also adds that "chromosomal disorders, such as Down syndrome, cannot be cured by gene therapy, nor can disorders resulting from complex interactions between many genes or between genes and environmental factors." Advocates of gene therapy contend that this form of treatment offers hope to the thousands of people whose diseases cannot be cured through current medical means. In 2000, researchers used gene therapy techniques to help mice with hemophilia produce high levels of the protein needed to restore and maintain the clotting property of blood. For advocates, knocking out this disease in the human population makes gene therapy—despite its limitations—a worthwhile pursuit.

Gene therapy is composed of two categories: somatic gene therapy and germ line gene therapy. In somatic gene therapy, therapeutic genes are introduced to the diseased cells of a patient in hopes that they will genetically alter them to function normally. In germ line gene therapy, therapeutic genes are introduced to reproductive cells (egg and sperm cells) to prevent the manifestation of a genetic disorder before the patient is born. This approach would alter the patient's genetic makeup and the genes he or she passes on to succeeding generations. Additionally, therapeutic genes can be introduced to cells in several ways. In ex vivo gene therapy, a patient's blood or bone marrow cells are removed and cultivated in a laboratory, exposed to a virus carrying therapeutic genes, and returned to the patient. In in vivo gene therapy, a virus or other particle carrying genes is inserted directly into the patient's body. The particle that carries genes to cells is known as a vector. Usually modified viruses are used as vectors in clinical trials, but the use of nonviral vectors, such as liposomes (microscopic fatty particles), are also under investigation. When genetic material is inserted without a vector, it is known as naked DNA.

The first human gene therapy clinical trial occurred in 1990, in which Ashanti DeSilva, then four years old, was treated for adenosine deaminase (ADA) deficiency, a rare genetic disorder that severely limits the functions of the immune system. Today, she leads a normal life and receives weekly injections of synthetic DNA to maintain her immune system. Some researchers herald the outcome of DeSilva's clinical trial as gene therapy's first success story, spurring interest and support for gene therapy research in the 1990s. However, hundreds of unsuccessful gene therapy clinical trials followed thereafter, dimming the initial optimism. But it was the death of a young patient that subjected gene therapy research to intense scrutiny. On September 17, 1999, eighteen-year-old Jesse Gelsinger died during a gene therapy clinical trial for ornithine transcarbamylase (OTC) deficiency, a rare metabolic disorder that is marked by dangerous levels of ammonia in the bloodstream. Although his condition was nonfatal and was controlled by a strict diet and regimen of drugs, Gelsinger volunteered to participate in an experimental treatment for a deadly type of OTC deficiency in babies at the University of Pennsylvania. He died after a vector injected into his liver triggered an immune response that led to multiple organ failure. The vector used to deliver therapeutic genes was a modified cold virus.

Immediately after Gelsinger's death, the Food and Drug Administration (FDA) froze all gene therapy clinical trials at the University of Pennsylvania and those under way at several other institutions. In addition, the prodecure of informed consent for clinical trial volunteers at the University of Pennsylvania was under fire. Some criticized the university for not thoroughly advising Gelsinger of the risks associated with the experiment in which he participated. Furthermore, an inquiry conducted by the National Institutes of Health (NIH) alleged that more than 650 adverse reactions in gene therapy trials were not immediately reported. In response, the FDA and the NIH took several steps to toughen the regulation of federally funded gene therapy research. They launched two initiatives in March 2000—the Gene Therapy Clinical Trial Monitoring Plan and the Gene Transfer Safety Symposia—to strengthen the oversight of gene therapy clinical trials and foster communication between gene therapy researchers. The FDA also conducted random investigations of seventy gene therapy clinical trials across the United States. Moreover, legislation to impose monetary penalties for the violation of clinical trial requirements (up to $250,000 per researcher and $1 million per institution) was drafted the same year. As of December 2003, such legislation had not been passed by Congress.

Proponents believe that increased regulation of gene therapy research is beneficial because it protects the safety of patients who volunteer for gene therapy clinical trials. According to Faith Lagay, a senior research associate at the American Medical Association, "We must better train (and perhaps certify) investigators to select, inform, and protect subjects in clinical trials" because "gene therapy illuminate[s] the weaknesses and cracks in our ability to monitor and enforce procedures for protecting human subjects and preventing their exploitation for science or commerce." Kathryn Zoon, director of the FDA's Center for Biologics Evaluation and Research, adds that monetary penalties "will give added assurances" to gene therapy patients that researchers and institutions are adhering to gene therapy research guidelines.

However, detractors of applying more restrictions to gene therapy research argue that it will needlessly delay its progress. Some assert that gene therapy is unfairly being singled out because other types of clinical trials expose their subjects to similar risks. The late Laura Raines, senior vice president of the Genzyme Corporation, a biotechnology company, claimed that "creating special rules exclusively for gene therapy research risks stigmatizing a product class for which the risks appear to be comparable to other types of products." Additionally, regarding the use of monetary penalties for regulation, Pamela Zeitlin, associate director of the Pediatric General Clinical Research Center at Johns Hopkins Hospital, suggests that they "would be very discouraging" for young researchers who contemplate joining gene therapy research, a field that urgently needs new recruits.

As of December 2003, gene therapy treatments are still experimental and have not yet been approved for any clinical use by the FDA. Although research pushed on after the tragic death of Jesse Gelsinger, the controversy surrounding it has not abated. In 2003 the field suffered another setback: Two French boys who were successfully treated for severe combined immunodeficiency in gene therapy clinical trials developed leukemia as a result of their treatment. In *At Issue: Gene Therapy*, the authors explore the benefits and risks involved in this young field of research as well as the significant implications gene therapy will have on human health if it becomes an acceptable form of treatment.

1

Gene Therapy: An Overview

Human Genome Management Information System

The Human Genome Management Information System (HGMIS) is part of the U.S. Department of Energy's Office of Biological and Environmental Research. HGMIS provides technical assistance and information to various research groups in hopes of making genome science accessible to a diverse audience.

Genes are the basic units of heredity. They are encoded to make proteins, which perform many life functions and compose most cellular structures. When genes are altered so that proteins do not function normally, genetic disorders may result. Attempting to correct genetic disorders in humans by replacing damaged or altered genes in a chromosome with normal genes is a technique called gene therapy. In most gene therapy cases, normal genes are introduced to the damaged region of the chromosomes through a "vector," usually a virus that has been manipulated to carry the normal genes. Either somatic (adult) or germ line (egg and sperm) cells may be manipulated through gene therapy. (In the latter, the new alterations are inherited by future generations.) Currently, gene therapy is still in the experimental stages and not yet approved to treat human disease by the U.S. Food and Drug Administration.

What is gene therapy?

Genes, which are carried on chromosomes, are the basic physical and functional units of heredity. Genes are specific sequences of bases that encode instructions on how to make proteins. Although genes get a lot of attention, it's the proteins that perform most life functions and even make up the majority of cellular structures. When genes are altered so that the encoded proteins are unable to carry out their normal functions, genetic disorders can result.

Gene therapy is a technique for correcting defective genes responsible for disease development. Researchers may use one of several approaches for correcting faulty genes:

- A normal gene may be inserted into a nonspecific location within

Human Genome Management Information System, "Gene Therapy," www.ornl.gov, 2003.

the genome to replace a nonfunctional gene. This approach is most common.
- An abnormal gene could be swapped for a normal gene through homologous recombination.
- The abnormal gene could be repaired through selective reverse mutation, which returns the gene to its normal function.
- The regulation (the degree to which a gene is turned on or off) of a particular gene could be altered.

How does gene therapy work?

In most gene therapy studies, a "normal" gene is inserted into the genome to replace an "abnormal," disease-causing gene. A carrier molecule called a vector must be used to deliver the therapeutic gene to the patient's target cells. Currently, the most common vector is a virus that has been genetically altered to carry normal human DNA. Viruses have evolved a way of encapsulating and delivering their genes to human cells in a pathogenic manner. Scientists have tried to take advantage of this capability and manipulate the virus genome to remove disease-causing genes and insert therapeutic genes.

Gene therapy is a technique for correcting defective genes responsible for disease development.

Target cells such as the patient's liver or lung cells are infected with the viral vector. The vector then unloads its genetic material containing the therapeutic human gene into the target cell. The generation of a functional protein product from the therapeutic gene restores the target cell to a normal state. . . .

Some of the different types of viruses used as gene therapy vectors:
- Retroviruses: A class of viruses that can create double-stranded DNA copies of their RNA genomes. These copies of its genome can be integrated into the chromosomes of host cells. Human immunodeficiency virus (HIV) is a retrovirus.
- Adenoviruses: A class of viruses with double-stranded DNA genomes that cause respiratory, intestinal, and eye infections in humans. The virus that causes the common cold is an adenovirus.
- Adeno-associated viruses: A class of small, single-stranded DNA viruses that can insert their genetic material at a specific site on chromosome 19.
- Herpes simplex viruses: A class of double-stranded DNA viruses that infect a particular cell type, neurons. Herpes simplex virus type 1 is a common human pathogen that causes cold sores.

Nonviral options

Besides virus-mediated gene-delivery systems, there are several nonviral options for gene delivery. The simplest method is the direct introduction of therapeutic DNA into target cells. This approach is limited in its application because it can be used only with certain tissues and requires large amounts of DNA.

Another nonviral approach involves the creation of an artificial lipid sphere with an aqueous core. This liposome, which carries the therapeutic DNA, is capable of passing the DNA through the target cell's membrane.

Therapeutic DNA also can get inside target cells by chemically linking the DNA to a molecule that will bind to special cell receptors. Once bound to these receptors, the therapeutic DNA constructs are engulfed by the cell membrane and passed into the interior of the target cell. This delivery system tends to be less effective than other options.

Researchers also are experimenting with introducing a 47th (artificial human) chromosome into target cells. This chromosome would exist autonomously alongside the standard 46—not affecting their workings or causing any mutations. It would be a large vector capable of carrying substantial amounts of genetic code, and scientists anticipate that, because of its construction and autonomy, the body's immune systems would not attack it. A problem with this potential method is the difficulty in delivering such a large molecule to the nucleus of a target cell.

The current status of gene therapy research

The Food and Drug Administration (FDA) has not yet approved any human gene therapy product for sale. Current gene therapy is experimental and has not proven very successful in clinical trials. Little progress has been made since the first gene therapy clinical trial began in 1990. In 1999, gene therapy suffered a major setback with the death of 18-year-old Jesse Gelsinger. Jesse was participating in a gene therapy trial for ornithine transcarboxylase deficiency (OTCD). He died from multiple organ failures 4 days after starting the treatment. His death is believed to have been triggered by a severe immune response to the adenovirus carrier.

Conditions or disorders that arise from mutations in a single gene are the best candidates for gene therapy.

Another major blow came in January 2003, when the FDA placed a temporary halt on all gene therapy trials using retroviral vectors in blood stem cells. FDA took this action after it learned that a second child treated in a French gene therapy trial had developed a leukemia-like condition. Both this child and another who had developed a similar condition in August 2002 had been successfully treated by gene therapy for X-linked severe combined immunodeficiency disease (X-SCID), also known as "bubble baby syndrome."

FDA's Biological Response Modifiers Advisory Committee (BRMAC) met at the end of February 2003 to discuss possible measures that could allow a number of retroviral gene therapy trials for treatment of life-threatening diseases to proceed with appropriate safeguards. FDA has yet to make a decision based on the discussions and advice of the BRMAC meeting.

Obstacles to effective treatment

- Short-lived nature of gene therapy: Before gene therapy can become a permanent cure for any condition, the therapeutic DNA introduced into target cells must remain functional and the cells containing the therapeutic DNA must be long-lived and stable. Problems with integrating therapeutic DNA into the genome and the rapidly dividing nature of many cells prevent gene therapy from achieving any long-term benefits. Patients will have to undergo multiple rounds of gene therapy.
- Immune response: Any time a foreign object is introduced into human tissues, the immune system is designed to attack the invader. The risk of stimulating the immune system in a way that reduces gene therapy effectiveness is always a potential risk. Furthermore, the immune system's enhanced response to invaders it has seen before makes it difficult for gene therapy to be repeated in patients.
- Problems with viral vectors: Viruses, while the carrier of choice in most gene therapy studies, present a variety of potential problems to the patient—toxicity, immune and inflammatory responses, and gene control and targeting issues. In addition, there is always the fear that the viral vector, once inside the patient, may recover its ability to cause disease.
- Multigene disorders: Conditions or disorders that arise from mutations in a single gene are the best candidates for gene therapy. Unfortunately, some of the most commonly occurring disorders, such as heart disease, high blood pressure, Alzheimer's disease, arthritis, and diabetes, are caused by the combined effects of variations in many genes. Multigene or multifactorial disorders such as these would be especially difficult to treat effectively using gene therapy. . . .

Recent developments

- University of California, Los Angeles, research team gets genes into the brain using liposomes coated in a polymer called polyethylene glycol (PEG). The transfer of genes into the brain is a significant achievement because viral vectors are too big to get across the "blood-brain barrier." This method has potential for treating Parkinson's disease. . . .
- RNA interference or gene silencing may be a new way to treat Huntington's. Short pieces of double-stranded RNA (short, interfering RNAs or siRNAs) are used by cells to degrade RNA of a particular sequence. If a siRNA is designed to match the RNA copied from a faulty gene, then the abnormal protein product of that gene will not be produced. . . .
- New gene therapy approach repairs errors in messenger RNA derived from defective genes. [The] technique has potential to treat the blood disorder thalassaemia, cystic fibrosis, and some cancers. . . .
- Gene therapy for treating children with X-SCID (severe combined immunodeficiency) or the "bubble boy" disease is stopped in France when the treatment causes leukemia in one of the patients. . . .
- Researchers at Case Western Reserve University and Copernicus

Therapeutics are able to create tiny liposomes 25 nanometers across that can carry therapeutic DNA through pores in the nuclear membrane. . . .

- Sickle cell is successfully treated in mice. . . .

Ethical considerations

- What is normal and what is a disability or disorder, and who decides?
- Are disabilities diseases? Do they need to be cured or prevented?
- Does searching for a cure demean the lives of individuals presently affected by disabilities?
- Is somatic gene therapy (which is done in the adult cells of persons known to have the disease) more or less ethical than germline gene therapy (which is done in egg and sperm cells and prevents the trait from being passed on to further generations)? In cases of somatic gene therapy, the procedure may have to be repeated in future generations.
- Preliminary attempts at gene therapy are exorbitantly expensive. Who will have access to these therapies? Who will pay for their use?

2

Gene Therapy Will Improve Human Health

Helen Phillips

Helen Phillips is a writer for Nature, *a biological and physical sciences magazine.*

Medical procedures that attempt to cure diseases on a genetic level are known as gene therapy. Since the first gene therapy clinical trial in 1990, hundreds of unsuccessful trials have followed, including the death of eighteen-year-old Jesse Gelsinger during a 1999 attempt to cure his ornithine transcarboxylase deficiency. In addition, in 2002, a French boy thought to have been cured of his rare immunodeficiency disorder developed leukemia as a result of gene therapy. These major setbacks, however, have challenged researchers to reevalute their procedures, ethics, and goals and carefully study their shortfalls for the development of better, safer techniques. Since gene therapy is a new procedure, only time is needed to fully realize its potential benefits. When the obstacles and hazards of gene therapy are surmounted, its potential to cure a wide range of life threatening and debilitating diseases will be a great asset to medical care and treatment.

"Great medical hope", "lethal experiment", "up-and-coming treatment". . . . In turns, gene therapy has been called each of these. So when French researchers announced [in October 2002] that a young boy given a therapeutic gene had developed leukaemia, there was a sense of resignation among researchers.

The knocks of gene therapy

Resignation but not despair. After the knocks that gene therapy has received [since 1990], researchers in the field are taking the news in their stride. Far from triggering a bout of hand-wringing and self-flagellation, the child's illness has confirmed gene therapists' understanding of the dangers of the treatment, and reinforced the recognised need to find better ways to deliver genes. It's too early to tell precisely how this case will

Helen Phillips, "Genes Can Come True," *New Scientist*, November 30, 2002, p. 30. Copyright © 2002 by Reed Business Information UK. Reproduced by permission.

affect the future of gene therapy but there is a feeling that the field will emerge stronger than ever.

Gene therapy hasn't seen this kind of optimism since the first trial in 1990 for an immune deficiency disorder. The experts thought that for diseases caused by single inherited mutations, such as cystic fibrosis, muscular dystrophy and thalassaemia, all they needed to do was add a healthy version of that faulty gene. And they would deliver the genes by co-opting the natural talents of viruses, which entwine their own DNA with ours. "The strategy was so simple and beautiful that we all got so excited," remembers Savio Woo of the Mount Sinai School of Medicine in New York, a former president of the American Society of Gene Therapy. The man who first proposed the idea of gene therapy, W. French Anderson of the University of Southern California School of Medicine, was equally beguiled. "We thought it would immediately translate into cures," he says.

But it didn't. Hundreds of unsuccessful trials followed, and experts began to doubt they'd ever solve the huge technical problems gene therapy was throwing at them. The field was harshly criticised by government agencies. Then disaster struck. When a young volunteer, Jesse Gelsinger, died inexplicably after a massive immune reaction in a trial in 1999, many thought the field would fold entirely.

A wholesale reappraisal

Instead, Gelsinger's death led to a wholesale reappraisal of the way gene therapy was conducted—its procedures, ethics and goals. A new realism emerged. And in September 1999, as this catharsis took place in the US, a group of researchers across the Atlantic at the Necker Hospital for Sick Children in Paris, led by Alain Fischer and Marina Cavazzano-Calvo, were notching up gene therapy's first cure, for another immune deficiency called X-SCID. At last somebody had shown that gene therapy could work.

Which is why it came as such a blow in October [2002] when the same team announced that one of its young patients had developed leukaemia, almost certainly triggered by the virus used to insert the gene into the boy. Immediately, questions were asked about whether all gene therapy trials should be stopped until safety could be assured. It certainly looked possible as regulatory agencies around the world halted all similar trials. The US froze recruitment for a forthcoming trial, while Germany and Italy suspended trials that used viruses similar to the one employed in France. Only Britain decided to continue testing the therapy, but with extra warnings and monitoring.

There is a feeling that the field [of gene therapy] will emerge stronger than ever.

But a couple of weeks later, having had a chance to discuss the case, both the US advisory committee to the Federal Drug Administration and the European Society of Gene Therapy recommended lifting these bans. Instead, researchers were urged to consider carefully the risks and potential gains for each disease being treated.

A risk worth taking

The reasoning was that alternatives to using viruses to shuttle genes into cells are years away from the clinic and the children involved in Fischer's trial almost certainly had no more than a year to live. So with no alternative treatment, leukaemia could be a risk worth taking. "It's not so much a setback for gene therapy," says Richard Mulligan of Harvard University, who is a member of the FDA advisory committee. "In my mind, it has eroded the risk-benefit ratio."

The list of gene therapies that are showing early signs of success is too impressive to give up.

Although observers were quick to draw parallels between the French boy and Gelsinger, there are crucial differences. Gelsinger's death was unpredicted, unexpected and to this day unexplained. Cancer, on the other hand, has always been the most obvious risk in the back of gene therapists' minds. Insert a gene into human cells and it could disrupt their normal functioning. Nobody knows how big this risk might be. But the child's case, tragic as it is, simply confirms a suspected problem. The challenge now is to work out how great the risk of cancer actually is, and how to deal with it.

Fischer's group has already discovered an impressive amount about the boy's leukaemia, which has helped to contain the damage. The boy was born in 1999 with a severe combined immune deficiency known as X-SCID. This is caused by mutation of an X-chromosome gene known as c, and it prevents two types of white blood cells, T cells and natural killer cells, from developing. With no defence against infection, sufferers usually die in their first year unless a bone marrow donor can be found.

Fischer and his colleagues collected stem cells from the boy's bone marrow and infected them in the lab with a retrovirus engineered to carry a healthy copy of the c gene. Although Fischer thinks that only about 50 of the treated cells picked up a working copy of the new gene, the fact that they were stem cells—precursors that can divide over and over to produce new blood cells—meant that when they were returned to the infant, they generated all the immune cells he needed.

The downside

By May [2002], Fischer's team had treated 11 children. All were doing well, living normal lives. Similar success was reported at London's Great Ormond Street Hospital. But in August [2002], the boy—Fischer's fourth patient—caught chicken pox. His T cell count stayed high after the infection and within a month it rocketed. It wasn't long before Fischer and his team concluded that the boy had a form of leukaemia.

They immediately suspected there might be a problem with the location of the therapeutic gene. Retroviruses, including the one used by the French researchers, integrate themselves into their host's DNA. The advantages are that the added gene can stay active for the lifetime of the cell, and when the cell divides, each daughter cell inherits the gene too. The downside is that it's not easy to control where the virus will end up.

And this seems to have been the problem. Fischer's analysis shows that in at least one bone marrow cell, the retrovirus inserted itself and the therapeutic gene into a regulatory region of a gene on chromosome 11 called Lmo2. It looks as though when the new gene became active, so did Lmo2. That might not be so bad except that Lmo2 is a cancer-causing "oncogene". Lmo2 could have triggered rapid cell division, producing a host of identical T cells—leukaemia.

The big question is how often will this type of thing happen? There are about 3 billion places the virus could have landed, and only around 300 oncogenes, so statistically it's unlikely that the virus would hit a danger spot. Some viruses, however, are known to prefer certain spots to lodge in, and if one of these is in or near an oncogene it would increase the odds of cancer.

About two-thirds of today's gene therapy trials are aimed at cancer and a handful have reached the large-scale studies that precede official approval.

Yet until now, cancer has only been seen in one animal test of gene therapy and never in human trials, leading researchers to suspect it poses only a minuscule risk. But Mulligan speculates that it could crop up more and more as we move into bigger trials. Early-stage clinical trials, designed for assessing safety rather than effectiveness, may have looked satisfactory only because the gene transfer was not very efficient, he points out. Quite simply, they seemed safe because genes weren't infecting many cells. The real risks will only appear during later-stage trials designed to test the efficiency of treatments.

Genetic differences

One factor that affects the level of risk may well be genetic differences between people which make some patients more susceptible to disease than others. Environmental factors, too, could make a difference. In the French case, chicken pox and a family history of cancer are just two variables to take into account. "We want to pass a clear message to the public that the treatment was to blame," says Mulligan. But there could be other factors behind the boy's leukaemia which will make it harder to determine the precise risk to others.

The treatment procedure itself may have avoided some potential hazards. Removing the bone marrow stem cells from the body before infecting them with the retrovirus eliminated the danger of an acute reaction to the virus. It also solved the problem of inserting the virus into the right cells, because they were the only ones in the dish. However, the flip side is that treating only a few stem cells and using them to repopulate the whole immune system amplifies a small risk into a big problem, says Mark Kay of Stanford University in California, who chaired the European Gene Therapy Society debate on the case. Precisely because the treatment is selecting for proliferating cells, it may be uniquely risky.

Theoretically, any virus that integrates its own genetic material into human DNA carries a similar risk, and many other therapies rely on such vectors. Indeed there is no point in giving up on integrating viruses in favour of those that remain loose, warns Alan Kingsman, CEO of British firm Oxford Biomedica, which researches and produces viral vectors. They don't stick around long enough to help in most conditions and almost any virus chosen to deliver DNA into cells will leave its traces. But at least the only unknown with integrating viruses is where those genes will land, not the order or number of them.

The realistic choice

Kingsman also says that Fischer's virus is now fairly old technology. Vectors can now be engineered to inactivate the signals that might, in rare cases, switch on an oncogene. In any case, Kingsman argues, viruses are still the only realistic choice at present, since other methods are just "horribly inefficient".

Other researchers think that viruses will never be part of optimal therapy because they are too complicated and costly. Researchers have tried alternatives such as encapsulating therapeutic genes in fatty globules called liposomes, or in the circular chromosomes called plasmids, which are found in bacteria. It's also possible to inject "naked" DNA straight into a target tissue. But all these processes are fairly hit-and-miss, and still incredibly inefficient.

At Stanford University, however, Michele Calos and her colleagues improved the efficiency of one of these techniques and developed a way to control exactly where the gene lands. They have teamed up plasmids containing a therapeutic gene with other plasmids containing a gene for a bacterial "integrase" enzyme. Integrases tend to cut the host DNA only at specific points and chaperone the gene into those spots.

At the moment, Calos and her colleagues have tested this approach only in animals and cultured human skin cells. They have a good idea where the integrase places the genes in mouse chromosomes—just two locations—and they are now searching for the sites favoured in human DNA. Each tissue will be different, so it won't be a quick job, but if the insertion points favoured in a particular tissue don't look safe and suitable, it is possible to evolve new integrases in the lab to do a better job.

Even chronic pain may be treated with gene therapy.

The researchers have used their idea to insert the gene for factor IX— a clotting agent that is missing in one type of haemophilia—into mouse liver cells, and with reasonable efficiency. While it will be at least a year before the group has enough data to start human trials, Calos is very optimistic about the future.

For now, though, cancer remains a risk. This will inevitably mean that the conditions tackled first will be life-threatening ones, where patients have little other hope. But researchers are trying to put the leukaemia case in perspective. Drug treatments carry risks, as do bone marrow transplants.

And the list of gene therapies that are showing early signs of success is too impressive to give up. To this day, SCID is the only disease that's been cured by gene therapy. Now there is hope for conditions including heart disease, cancer, Alzheimer's, Parkinson's, AIDS and even chronic pain.

Only as trials go on and researchers get a better idea of the cancer risk will we decide which treatments become routine. If the risks are high, we'll have to wait for alternative technologies, rather than giving up. By the time you or I suffer one of these diseases, we may well be offered a gene therapy.

The next wave

One of the strongest contenders to win the first gene therapy licence is a treatment for haemophilia B. Sufferers lack the gene for factor IX, a crucial agent in blood clotting. A team led by Mark Kay of Stanford University, California, is using parvoviruses to insert the missing gene into liver cells. The cells generate the factor, removing the need for daily injections. The team hopes to reveal the results of their latest trial in December [2002]. Richard Mulligan of Harvard University believes this approach should soon yield positive results.

Haemophilia is one of a group of relatively common single-gene disorders that were originally expected to succumb swiftly to the powers of gene therapy. But it is the only one to show any real promise in the clinic. Others, such as cystic fibrosis and muscular dystrophy, have proved more difficult to treat. Instead, gene therapy's main targets have changed radically, says Savio Woo of Mount Sinai School of Medicine in New York. "We have begun to think about using genes for all kinds of medicine," he says.

Cancer treatments

Cancer treatments are also in the running for licences. About two-thirds of today's gene therapy trials are aimed at cancer and a handful have reached the large-scale studies that precede official approval. The approaches are varied. Some use genetically modified viral vectors to prime the immune system to attack cancer cells. Others employ viruses to carry suicide genes into tumour cells. Researchers have also developed viruses that only replicate in cancer cells, so killing them while leaving healthy tissues untouched. And one of the large trials under way tackles head and neck carcinoma by replacing a faulty tumour-suppressing gene called p53. "It's possible that cancer will be the second cure," says gene therapy pioneer W. French Anderson of the University of Southern California.

Specially engineered HIV may eventually be recruited to help control HIV-1 infection. Researchers from the National Human Genome Research Institute in Bethesda, Maryland, have produced an apparently harmless form of the virus that seems to outcompete the disease strain. It grows more rapidly and uses up the limited supply of raw materials needed to form infectious virus particles.

The range of novel ideas

The range of novel ideas for gene therapies is staggering—genes for nerve growth factor in Alzheimer's patients, different growth factors for Parkinson's, genes for cell surface proteins that reverse male sterility, blood vessel growth factors for heart disease, and genes that control the immune response to block autoimmune diseases. All are being tried.

Even chronic pain may be treated with gene therapy. David Fink of the University of Pittsburgh and the VA Pittsburgh Healthcare System has exploited the way the herpes simplex virus travels along the tendril-like axons of sensory nerves and hides in the nerve cell bodies near the spinal cord. He's engineered the virus to carry a gene for the body's natural painkiller, enkephalin, directly to those nerves causing pain. Enkephalin is far too short-lived in the body to give as a drug, but manufacturing it directly in the cells that transmit pain signals is an exciting prospect for treating the pain caused by nerve damage, diabetes and cancer.

Despite its tragedies and setbacks, gene therapy is getting there, says Anderson. "It takes 10 years to get a drug through to approval," he adds. "Gene therapy is basically a new medicine. We're just 13 years into it. By the time we're 15 years into it we'll start to see approved treatments."

3

Gene Therapy May Only Benefit the Wealthy

Mohamed Larbi Bouguerra

Mohamed Larbi Bouguerra is the former director of the National Institute for Scientific and Technical Research in Tunisia. He has written several books about the negative effects of new science and technology on the third world and the environment.

Gene therapy and research, which aims to treat genetic disorders and diseases by replacing faulty genes, is expected to serve as the basis of medical advances in the twenty-first century. However, this technology is very expensive and likely to be inaccessible to third-world people. Also, because many impoverished countries are not on the forefront of genetic research, they are not included in current debates regarding the ethics and implementation of gene therapy. But their participation is vital because alterations in the human gene pool could unexpectedly have adverse effects on third-world populations. Hence, measures that make gene therapy technologies more accessible to developing nations and encourage worldwide participation in the genetics debate should be enforced.

New drugs and treatments based on genetic research are set to widen existing disparities in access to medical treatment.

Medicine in the 21st century is likely to be based on genetics.

The decipherment of DNA being done as part of the international Human Genome Project will probably unleash a flood of applications which will make it possible to improve the physical condition of human beings at a time when many experts are saying the limits of "conventional" medical care have been reached.

Over the next two decades, gene therapy, immunology and cell culture enabling production of totally uncontaminated blood (for people with leukemia, for example) are expected to make great strides. The availability of a range of prenatal tests designed to spot genetic anomalies in embryos will boost the development of genetic counselling services.

But who will benefit from all this? Right from the start of the Human

Genome Project in 1990, James D. Watson, one of the discoverers of DNA's double helix structure, has campaigned for this great project to stay in the public domain. "The world's nations," he has said, "must realize that the human genome belongs to everybody on the planet and not to individual countries."

Most discoveries and new treatments have come out of laboratories in the rich countries, but people from countries of the South have also contributed their brain power and hard work. It was an Indonesian, Joe-Hin Tjio, who proved, in 1956 in Sweden, that human beings have 46 chromosomes. In 1968, in the United States, Indian Nobel Prize–winner Har Gobind Khorana became the first person to synthesize a human gene.

Genetic data and the poor

Analysis of the DNA of some indigenous peoples has yielded valuable genetic data for scientists who have subsequently declared discoveries made on the basis of it to be their own intellectual property. One such population study, for example, identified the genes of a man of the Hagahai tribe (Papua New Guinea) which provide immunity to the leukemia virus HTLV.

But owing to lack of resources and political will, many poor countries have trouble putting together a serious policy on science which would reduce their total dependence on rich countries and enable them to work out research priorities. Some of these countries however have human expertise and facilities which allow them to contribute to work on DNA sequencing [identifying genes by determining the order of bases in DNA].

India, for example, has six laboratories where this can be done, all of them linked to the Hyderabad Centre for Cellular and Molecular Biology. Some Indian specialists would have preferred to sequence the DNA of pathological organisms (microbes, mosquitoes, contaminants, etc.) which are common in India, rather than that of human beings chosen at random. In this way, it would have been easier, they say, to develop applications that would be immediately useful to people in India. This is an ongoing debate because nobody can guarantee that analysis of the human genome will lead to medical treatment for people in the countries of the South, where there are fewer profits to be made than in the North.

"Medical apartheid"

The other big question is that of access to the new forms of treatment. Even in rich countries, where public spending on health is being cut, these therapies will be very expensive—at least to begin with—and are likely to trace a new frontier between the well-off and the rest of the population. So it is not very likely that treatment of this kind will really reach the people of the third world. The countries of the South are also light-years away from the new horizons of medicine because they lack basic health facilities and trained health workers.

Should some countries, or even whole regions like Africa south of the Sahara, not be part of today's debate about bioethics just because they might be excluded from the benefits of tomorrow's medicine? The answer is no. First, because their inhabitants are sometimes directly involved. The elimination, for example, of some "harmful" genes through germline therapy—if it were ever applied on a world-wide scale—could be very harmful to them. In accordance with the phenomenon known as pleotropy, a single gene can control several characteristics. Thus the recessive gene of cystic fibrosis may play a part in fighting cholera.

Sickle-cell anaemia (an abnormal form of the red pigment of the blood, haemoglobin) affords some protection against the deadly form of the malaria parasite, Plasmodium falciparum. If the gene that triggers this condition is eliminated, do we risk seeing even more cases of malaria? This is a very gloomy prospect in a world where malaria is already killing two million people a year and when none of the big pharmaceutical companies have invested money to look for a vaccine. More generally, the whole planet is concerned by the risk of reducing the genetic reserves available to future generations by altering or eliminating certain genes. Is it not presumptuous and dangerous to anticipate their needs when nobody yet knows what kind of environment they will be living in?

[Gene] therapies will be very expensive . . . and are likely to trace a new frontier between the well-off and the rest of the population.

No one on the planet should be excluded from discussions about matters affecting human existence, which are of course of universal concern. Bioethics is about the absolute, intrinsic worth of every individual— the very essence of human life. The lines it draws between what is possible and what is acceptable should be worked out with the involvement of all the world's cultures, even if they are minority or dominated cultures.

What's more, bioethics again poses the urgent question of solidarity among human beings in the face of illness. When genetic medicine no longer just serves the rich but also enhances their lifestyle, prolongs their lives or enables them to produce children with specific characteristics, can we deny the people of poor countries the benefits of knowledge which would free them from the scourge of debilitating parasitical diseases, of AIDS and of hereditary afflictions? How long could such "medical apartheid" fail to affect the consciences of those living in the countries of the North?

A treaty on the genome?

"In the framework of international co-operation, States should seek to encourage measures enabling: . . . the capacity of developing countries to carry out research on human biology and genetics, taking into consideration their specific problems, to be developed and strengthened . . . ; developing countries to benefit from the achievements of scientific and technological research so that their use in favour of economic and social progress can be to the benefit of all." Article 19 of the Universal Declaration on the Human Genome and Human Rights gives a new focus to "the rights of solidarity". The declaration, which is not legally binding, was adopted by the international community in 1997 after long negotiations within UNESCO's (United Nations Education, Scientific and Cultural Organization) International Bioethics Committee (IBC). The declaration also enshrines two major principles, explains Noëlle Lenoir, who chaired the IBC during the negotiations. First, rejection of biological determinism: human beings are not animals programmed by their genes. Second, a refusal to accept that genetics can provide justification for socially discriminatory or racist practices. "Human dignity" is the key expression in the text, which condemns reproductive cloning. This declaration adopted under UNESCO's auspices is today the only text of universal scope which specifically concerns bioethical issues. "But," Lenoir adds, "I feel that in the present context of globalization we should be moving towards a treaty," that signatory states would be bound to respect. Warning: turbulence ahead.

4

Gene Therapy Research Has Made Significant Advances

Josh P. Roberts

Josh P. Roberts is a freelance writer in Minneapolis, Minnesota.

Although advancements in gene therapy have not been as dramatic as the first clinical trial over a decade ago, in which four-year-old Ashanti DeSilva's deadly immunodeficiency was successfully treated in part by genetic intervention, research in the field is making progress. The death of eighteen-year-old gene therapy patient Jesse Gelsinger in 1999, and other setbacks in its short history, placed gene therapy research under increased scrutiny and stricter regulations. But such events overshadow the incremental advances gene therapy researchers are making today. In fact, a wide variety of trials are currently under way, including gene therapy experiments aimed at treating cardiovascular disease, HIV, and cancer, that are developing more efficient techniques of genetic intervention.

In 1990, three men—W. French Anderson, R. Michael Blaese, and Kenneth Culver—led a trial in which the genetically corrected adenosine deaminase (ADA) T cells, belonging to a 4-year-old girl, were returned to her. Today, the [young woman] is alive and well.

It took another decade or so for any accomplishments as dramatic as that first trial to be reported, due in part to a relatively empty toolkit. In April [2002], following a trial of gene therapy that occurred two years prior, French researchers announced that the immune systems of several children severely affected with X-linked severe combined immunodeficiency (SCID) were nearly normal, and that no supplementary therapies were involved. Other, less headline-grabbing reports also occurred, including work on curing fatal congenital diseases, reversing infertility in mice, treating patients with hemophilia, and combining different therapies with gene therapy.

The death of gene therapy?

The short history of gene therapy is like a roller coaster, with quick, adrenaline-creating ascents and tortured, heart-in-the-mouth descents.

Since gene therapy patient Jesse Gelsinger died in 1999 at the University of Pennsylvania, this field has been subjected to intense scrutiny, with new regulations established to police experiments and new protocols to follow. Moreover, media coverage of the slow, steady research progress— what Anderson calls routine—had taken a back seat to headlines that sensational setbacks such as Gelsinger's death have received.

But (with apologies to Mark Twain) the death of gene therapy has been greatly exaggerated. At least 2,000 labs are engaged in gene-therapy research worldwide; an Internet search of "gene therapy" in PubMed indicates that nearly twice as many papers were published in 2001 than five years earlier. At the National Institutes of Health, monies devoted to lab and clinical research increased 22% from $349.4 million in fiscal year 2001, to an estimated $427.4 million in 2003. Moreover, membership in the American Society of Gene Therapy (ASGT) has grown from 1,000 when it was founded in 1996, to 3,000 members in 2002. More than 600 gene therapy trials are ongoing worldwide. Of the 509 trials listed in the NIH's pilot Human Gene Transfer database (www4.od.nih.gov/oba/rac/ clinicaltrial.htm), about two-thirds are for the treatment of various forms and stages of cancer. A wide variety of other trials are ongoing as well, most of which fall under the headings of cardiovascular disease (46 trials), infectious disease—nearly all HIV-related (40 trials), and inherited autosomal recessive (44 trials).

At least 2,000 labs are engaged in gene-therapy research worldwide.

Recently, the press has begun paying more attention to some of the successes, printing headlines including "For Gene Therapy, a Humble Return," and "Gene Therapy Gives Heart Patients Hope." Reports of successfully treating the rare X-SCID—from INSERM [Institut National de la Santé et de la Recherche Médicale] in Paris as well as similar news from London's Great Ormond Street Hospital—were greeted worldwide with headlines such as "Gene Therapy Rids 'Bubble Boy' Disease."

"I think we should hold up the fact that you can actually do gene therapy and it really works. That's pretty exciting," says Fred Hutchinson Cancer Research Center virologist Dusty Miller, whose vectors have been used in numerous gene therapy trials. All is not glorious, however: No clinical trials have yet to complete the large Phase III trials necessary to win the Food and Drug Administration's (FDA) approval of therapeutics.

Ancient history

Clinical gene transfer had its official beginning when, in 1989, five patients with terminal melanoma were given autologous lymphocytes that had been "marked" ex vivo with a gene encoding resistance to the antibiotic G418. This study was designed primarily to trace the cells in the patients' bodies and to show the safety of gene transfer, and in that sense it was successful. No helper viruses were found, no reverse transcriptase activity was detected, no toxicity was experienced, and the transduced cells remained otherwise "normal."

That trial paved the way for the world's first sanctioned gene-therapy trial. In 1990, 4-year-old Ashanti de Silva's ADA T cells were genetically corrected and then returned to her. She is still on a low-dose regimen of intravenous PEG-ADA therapy, her immune system is now fully functional, with 20% to 25% of all her T cells containing the gene that was introduced by retroviral transfer in 1990, says Anderson, who is now at the University of Southern California.

"We should hold up the fact that you can actually do gene therapy and it really works."

At the time, ADA deficiency was "really the only disease that we could think of that we thought we had a shot of helping," recalls Blaese, former chief of NIH's Clinical Gene Therapy Branch. For one thing, transduction efficiencies were "terrible," he explains. Culver, now at Norvartis, comments that although relatively few T cells would become transduced, these had a tremendous selective advantage over their endogenous counterparts. As a bonus, the introduced ADA gene product helped endogenous T cells to thrive as well.

T cells can also be easily removed and isolated, then grown and expanded in culture. This is important because retroviral vectors, the only ones available at the time that were capable of stable transduction, could transfer genes only to dividing cells. And T cells can be reintroduced into their proper places in the body.

Steady progress

The recent advances witnessed in gene therapy reflect a large number of incremental steps in many areas, rather than one or two "great strides," Blaese and others note; these include more and better viruses. Anderson cites as an example the incremental improvement of culturing conditions, including growth factors, which allows for significantly greater transduction. What were once mystery factors in tissue culture supernatant, such as the T-cell growth factor IL-2, can now often be purchased off-the-shelf in purified form.

Anderson also points to scientists' hard-won ability to determine which vector will work best in which type of application. "There isn't going to be a 'magic vector' that is useful for every situation." Not only do many vector classes exist today—the big five are retroviral, adenoviral, adeno-associated viral (AAV), lentiviral, and nonviral, with many others being investigated as well—but the vectors themselves have improved, says Anderson, founder and editor of *Human Gene Therapy*.

Some improvements include "gutless" adenoviruses that allow for larger genes to be inserted and have less potential to evoke an immune reaction to the vector (which is thought to have contributed to Gelsinger's death). Viral genes are now generally supplied by the packaging cell in *trans* (as has been the case for most other vectors), giving added assurance against transferring active virus to the patient. And viral vectors can now be routinely pseudotyped to achieve a desired tropism. Investigators can now choose from a "whole toolkit of viruses," notes Miller.

Although many would disagree, Miller does not think that nonviral technologies, such as introducing naked or plasmid DNA by gene gun or liposomal transfer, are, at this point, very efficient for gene therapy. These technologies have no specific integration mechanism, and the gene tends not to persist in the target cell, he observes. "Nature has developed some pretty good tools. Viruses have figured out over billions of years what to do, and it will take a while for scientists to do the same."

However, one of the four early-stage clinical success stories cited by Anderson in a *Nature Medicine* commentary involved using naked plasmid DNA to induce angiogenesis in cardiovascular disease patients. Illustrating the toolkit concept, the other trials involved retroviral transfer (the X-SCID trial), AAV (used to treat hemophilia), and an oncolytic adenovirus.

Cancer and gene therapy

The latter trial, which Anderson calls the first successful Phase II trial for cancer, used the mutant adenovirus to specifically replicate in and lyse carcinomas that had lost the ability to make the tumor suppressor *p53*. The therapy was not effective on its own, but it did show a significant benefit when combined with standard chemotherapy.

Since the field's seminal days, oncologists have investigated gene therapy's potential to treat their patients. Most early lab and animal experiments—few actually made it to humans—were variations on one theme: Researchers tried to make tumors more immunogenic, partially because most of these did not require the long-term transduction of many cells. Most experiments have met with only limited success. Blaese cites the fundamental premise of inducing the immune system to fight the disease—rather than problems of gene transfer—as the major limitation of these early studies, but he does admit that "there are some studies along those lines showing some levels of efficacy."

Immunotherapy still dominates researchers' efforts to treat malignancies with gene therapy; investigators are using strategies ranging from returning gene-enhanced irradiated tumors to patients, to injecting tumor-specific antigen-engineered pox virus into muscle.

Two other heavily investigated oncological strategies are gene replacement—the same concept used in treating patients with SCID—and direct or indirect killing of the malignant cells. In the former, a working copy of a gene that has gone awry (for example, *p53*) is introduced into the tumor to recheck its growth. In the latter, researchers introduce a "suicide gene" such as the thymidine kinase gene of the herpes simplex virus, whose product will poison the cell on exposure to the antiviral drug gancyclovir.

On trial

The majority of new gene-therapy trials, says Anderson, involve either cancer or vaccines, and sometimes both. Initially, most of them involved ex vivo genetic manipulations followed by reintroduction of the gene-engineered tissue. Now, the tide is changing, Anderson notes, with vectors being delivered in situ and introduced directly into the target tissues.

But whether these new trials hold promise has yet to be seen. The vast majority of clinical trials have not made it to Phase II, and those dealing with rare genetic disorders such as ADA "never will," Anderson points out. The issue is money: the NIH, private foundations, or perhaps an academic institution, generally bankroll these trials. Says Anderson, "There's no money in it. No drug company is going to spend . . . $100 million to go through all the pivotal Phase III trials that are necessary for approval." Such trials themselves may eventually become standard-of-care, but without "that golden piece of paper from the FDA that says your NDA [new drug application] has been approved," Anderson explains, "it is still a trial."

Of course, pretrial work continues—the recent ASGT conference produced a record number of abstracts, Blaese says. One of these, from Miller's lab, describes investigations into treating patients with cystic fibrosis by introducing a functional cystic fibrosis transmembrane conductance regulator gene. That gene, with all of its regulatory elements, will not fit into a vector, necessitating use of the cDNA with shortened transcriptional elements. Even this does not easily fit into an AAV vector (which, he says, works best in the lung), so they developed a way to split the shortened gene into two AAV vectors and then allowed them to recombine in vivo. The lab is also trying to adapt an oncogenic sheep retrovirus, which replicates in the lung, for use in human gene therapy. "There is a new batch of vectors evaluated every year," Blaese notes.

Observers point to several new developments that hold great promise for the field: lentiviruses (such as HIV); transposons, which can stably introduce genetic material into quiescent cells; and continued improvements in nonviral delivery systems. Blaese also expects to see a lot of "transgenomic viruses," combining aspects of different viruses, as well as combinations of viral with nonviral vectors and strategies. The field is now vibrant and healthy, he says, "working its way through its growing problems."

5

The Advances of Gene Therapy Research Have Been Exaggerated

Angela Ryan

Angela Ryan studies molecular biology at King's College in London, England.

Clinical advancements in gene therapy research have been overstated. The so-called successes of gene therapy are merely anecdotal and have not demonstrated clinical efficacy, yet they are oversold by scientists seeking research funding. Furthermore, in 2000, gene therapy clinical trials were halted when it was learned that serious adverse effects caused by such trials—including infections, fever, and unexplained deaths—were not reported to the National Institutes of Health. Nonetheless, current gene therapy protocols often involve techniques that are ineffective and possibly dangerous, putting human subjects at risk of serious adverse effects or illness. Real progress and safety in gene therapy research cannot be achieved until the complexities of the human genome are carefully studied and fully understood.

[A pril 2001's] *New Scientist* reported the combination of two notorious killer viruses, HIV and Ebola, in an attempt to find an effective gene therapy vector for the treatment of cystic fibrosis. When this work was presented at a scientific meeting the audience laughed out loud.

Gene therapy is targeted at virtually every ill known to human beings, especially those inhabiting the first world, including pain relief, cosmetic hair replacement and muscle building. Massive investment has gone in but no clinical efficacy has ever been proven, despite anecdotal claims of success.

[In 2000] in the US, gene therapy clinical trials ground to a halt amid scandalous reports of deaths and conflicts of interest. The US National Institutes of Health (NIH) set up a special telephone hot line for victims that counted 652 cases of serious adverse events along with six unexplained deaths. Effects included high fevers, infections and severe changes in blood pressure, all of which went previously unreported to the NIH Recombinant DNA Advisory Committee (RAC). David Baltimore, Nobel laureate and president of Caltech, a gene therapy based biotech company, said "I disagree we've had any benefit from gene therapy trials so far, many of us are now asking, what the hell are we doing putting these things into people?"

Sir David Weatherall, Professor at the Institute of Molecular Medicine, University of Oxford, told The UK Royal Society discussion meeting on Social Responsibility in Science that "scientists have not made efforts to maintain an open and completely honest debate with the public about what they are doing. Part of the problem arose from over ambition or pressures to publish, to attract research funding".

Hype in the media

Misinformation has generated much hype in the media about the promises of gene therapy. One main problem identified by Weatherall is that, "many scientists working in the molecular sciences are not clinically trained, even though their work impinges more and more on human molecular pathology. They know a great deal about the technicalities of their field but nothing about the complexity of human beings and their diseases". Scientists have over-exaggerated their work, for newspapers don't like 'ifs' and 'buts'.

The US Food and Drug Administration (FDA) and the NIH responded to widespread concern about risks, especially after the 1999 death of teenager Jesse Gelsinger in a phase 1 clinical trial. Many laboratories were shut down, public meetings were held, reviews and investigations commissioned and administrative changes have been put in place to deal with the crisis. But the troubles run deep within the heartland of biomedical science, where the most important concern remains the issue of *safety*.

Gene therapy targets diseases based on the transfer of genetic material into an individual, rather than a drug. It uses genes as the therapeutic agent, and it is qualitatively very different from other forms of treatment. Despite the serious health risks involved, clinical trials have been underway since 1990. The recently released NIH 1995 report on gene therapy research documents a plethora of scientific and clinical risks associated with gene therapy, many of which have been highlighted independently in an ISIS [Institute of Science in Society] report.

Major technical problems

There are major technical problems with all aspects of gene therapy. Furthermore, few pre-clinical data have been published and toxicological evaluations are seldom found in the literature. The potential for generating new viruses, known as replication-competent viruses (RCV) needs to be thoroughly evaluated, particularly as genetically modified viruses are used in gene therapy. The spread of viral vectors to non-target tissues throughout the host is also a major safety concern. There is no way to predict the virulence or disease potential of recombinant viral vectors, and a case-by-case approach has to be applied. It has been shown, however, that viral vectors can induce toxic shock following administration.

The NIH expert panel found that all gene transfer vectors are ineffective and it is not understood how they interact with the host. Basic studies of disease pathology and physiology have not been done, which are critical for designing treatment. It is not possible to extrapolate from animal experiments to human studies. In the cases of cystic fibrosis, cancer and AIDS, animal models do not have the major manifestation of the disease in humans. Gene transfer frequency is extremely low and results of gene therapy protocols rely on qualitative rather that quantitative assessments of gene transfer and expression. There are no controls, and biochemical or disease endpoints are not defined.

The panel concluded "only a minority of clinical studies, illustrated by some gene marking experiments, have been designed to yield useful basic information" [as these at least track the fate of the genetic vector]. The report states that there is "concern at the overselling of results of laboratory and clinical studies by investigators and their sponsors, either academic, federal, or industrial, leading to the widespread perception that gene therapy is further developed and more successful that it actually is".

Purely speculative

In gene therapy, DNA is delivered, either by direct administration of viral vectors, or naked DNA, into the bloodstream or the tissues, or indirectly, through the introduction of cells that have first been genetically modified. In human studies, only somatic cells are the target of gene therapy, not germ cells (eggs and sperm), although germ line gene therapy is common practice in animals. Four main types of disease are targeted; single-gene inherited disorders, multi-factorial disorders, cancer and infectious diseases.

Gene therapy is targeted at virtually every ill known to human beings.

Single-gene inherited disorders occur infrequently in populations. They are chronic conditions associated with the loss of function in a gene and relevant protein. Such single-gene disorders include sickle cell anemia, hemophilia, inherited immune deficiencies, hyper-cholesterolemia and cystic fibrosis. Gene therapy aims to replace the mutant gene with its normal counterpart. The NIH panel found major problems with access to relevant cell types as well as assessing the total fraction of cells in a tissue that need to be corrected. It may not be technically possible to achieve the right level of gene expression required for correction, nor regulating the expression of the gene after it is transferred.

Multi-factorial disorders, like coronary heart disease or diabetes, involve many genes, not to mention environmental factors. The aim of gene therapy is to reverse or retard disease processes at the cellular level. The NIH panel pointed out that it is "not known how specific gene products influence cellular physiology" and therefore only purely speculative strategies have been proposed and tested.

Concerns over safety

Last year, the American Heart Association (AHA) expert panel on clinical trials of gene therapy in coronary angiogenesis found gene therapy to be unsatisfactory, especially in comparison to conventional treatments, and expressed concerns over safety.

Gene therapy for coronary angiogenesis involves the delivery of growth factor genes into the heart to stimulate blood vessels to grow. But the Heart Association stated "no process-specific stimuli or growth factor has ever been identified", and "re-growth of blood vessels is a complex process that involves multiple levels of stimulators, inhibitors and modulators". Therefore, for a single growth factor to work, "an entire self-propagating cascade or proliferative, migratory, chemotactic and imflammatory processes must be initiated". They leveled strong criticism to suggest that gene therapists aren't even using the right genes.

Scientists have over-exaggerated their work [in gene therapy], for newspapers don't like 'ifs' and 'buts'.

The Heart Association is also concerned over the mode of delivery and the 'optimal dose schedule', which they said "is unknown". Gene therapy is very variable in the levels of the proteins produced and the duration of expression. They cite one study in which earlier-generation adenovirus vectors persisted and caused dysregulation of a number of host genes. They state that "preclinical and clinical studies *should be preceded* by tissue distribution studies to define the myocardial uptake and retention or expression of growth factors" (author's emphasis).

Gene therapy vectors cause immune responses, which in turn cause inflammation and transgene silencing. Attempts to make vectors safer and more efficient result in longer-term transgene expression and the American Heart Association expressed concern about deleterious effects due to prolonged growth stimulation. They are also concerned about cancer, a known risk with all gene therapy protocols due to random insertion of transgenes into the cell's genome. The report states quite categorically "the necessary extent of cancer screening has not ever been defined".

Unsuccessful attempts

The NIH panel pointed out that in many cancers, the cancer causing gene is dominant and transferring a normal copy has no impact. The number of cells within a tumor is large, and the technology will only transfer genes to a subset of cells within a tumor mass. Furthermore, the mutation rate in cancer cells is very high, so the introduced gene itself may become mutated, its function inactivated, giving rise to more cancer cells. Finally, the complication of migrating cancer cells means the transfer of DNA is "not a feasible strategy".

More indirect gene therapy approaches have been considered for cancer, including the transfer of genes for cytokines or other immune modulatory factors, either outside or inside the body of the patient. This approach attempts to stimulate immune recognition not only of tumors but also cancer cells that have spread. Some of these strategies have shown promise in mouse models but none have demonstrated efficacy in humans.

A number of chronic infectious diseases have been targeted by gene therapy, HIV being the best studied. Efforts have focused in two areas; post-exposure vaccination and attempts to express genes in target cells that render HIV unable to infect or replicate. Other products have been developed and tested, including mutant proteins that inhibit virus replication, antisense RNA that blocks translation of HIV genes, ribozymes that break down HIV RNA, 'decoy' RNA that competes for binding of viral proteins and antibodies that prevent key HIV enzymes from functioning. All these strategies and more have been attempted, without success.

Naked DNA vaccines for HIV contain single HIV genes or combinations or HIV-1 early regulatory genes. Such HIV derived genes may recombine with other retroviral sequences, generating new strains. Viral sequences also integrate into the host genome, causing genetic damage.

Gene transfer vectors

Three main types of gene transfer vector systems are in use: DNA vectors (either naked or complexed with proteins or other molecules), RNA viruses (retroviruses), and DNA viruses (adenovirus, adenoassociated virus [AAV], herpesvirus, and poxvirus). However, none of the available vector systems are satisfactory.

The NIH report stated "the perceived advantages of each system have not been experimentally validated", and "the efficient introduction of these vectors into cells is likely to be a formidable obstacle to their use."

Retroviral vectors are used extensively, as the basic biology of retroviruses is the best understood of the vector systems. But they are very expensive and complicated to prepare and validate, often having a low titer and limited insert size. Gene transfer is limited to dividing cells and expression is difficult to control and stabilize. They insert randomly in the host chromosome, which causes genetic damage and means the introduced gene does not express in the same way as it would in a normal, healthy cell. They can also lead to the creation of new viruses.

Adenoviral vectors have been used in about 25% of active gene therapy trials. They contain many viral genes and have been shown to be highly immunogenic. They can enter most cell types, although the factors controlling this are poorly understood. They generate RCVs by recombination and cause genetic damage by random integration into the host genome. Patients with previous infection of natural adenovirus will mount immune responses to these vectors.

Worse than the disease

Teenager Jesse Gelsinger died three days after receiving a dose of adenoviral vectors. Within the first day, tests showed he had suffered liver injury and inappropriate blood coagulation. On the third day he had trouble breathing and his vital organs began to fail. He was taken off life support on the fourth day. The autopsy revealed further abnormalities. The researchers had concentrated the vector in the liver, infusing it directly through a catheter. But significant amounts of vector were found in the spleen, lymph nodes, bone marrow and other tissues and when analyzed, duplicate sequences not engineered in the original were discovered, revealing vector recombination. . . .

Viral coat proteins are also being used to help improve the uptake of viral vectors; this is known as pseudotyping. Retroviruses pseudotyped fuse with cells and do not use their normal receptors to gain entry into cells. They have a much broader host range than wildtype viruses and some are capable of infecting all organisms, showing no restriction for species infectivity. Such viral particles are potentially very dangerous and should not be released from contained use conditions. They may recombine with wild viruses and relays of horizontal gene transfer events could bring about the creation of a new viral zoonosis, causing a world pandemic. . . .

'Gene therapy' has been wildly premature. All the indications suggest this so called 'therapy' may be worse than 'disease'. Many scientists have pointed out that 'complexity' is the watchword in disease genetics. Even the apparent simplicity of single-gene disorders is clouded by the specter of modifier genes that can influence disease susceptibility, severity or progression. Genetic determinism is dead. Much careful work is required to tease apart the complexities of the range of factors that influence normal gene expression.

6

Germ Line Gene Therapy Will Improve Human Health

James D. Watson and Andrew Berry

James D. Watson, along with Francis Crick and Maurice Wilkins, won the Nobel Prize in Physiology or Medicine in 1962 for the discovery of the structure of DNA (deoxyribonucleic acid). Andrew Berry is a research associate at Harvard University's Museum of Comparative Zoology. Watson and Berry are coauthors of DNA: The Secret of Life, *from which the following excerpt is taken.*

In gene therapy, either somatic (adult) or germ (egg or sperm) cells are manipulated to correct the harmful effects of abnormal genes. The latter, known as germ line gene therapy, has not been attempted but shows the potential to prevent abnormal genes from being inherited by subsequent generations. Thus, when it becomes a viable and fail-safe option, germ line gene therapy should be used to protect individuals from arbitrary and cruel genetic disadvantages, such as crippling diseases and learning disabilities. In addition, it should be used to battle the AIDS epidemic by building human resistance to HIV. The only rational objection to germ line gene therapy lies in its safety, since one single error could deleteriously affect an individual's life and possibly his or her offspring. Nonetheless, for the sake of human health, society should draw upon the courage used to pioneer current medical procedures and explore the possibilities germ line manipulation may offer.

Even those who accept that the urge to improve the lot of others is part of human nature disagree on the best way to go about it. It is a perennial subject of social and political debate. The prevailing orthodoxy holds that the best way we can help our fellow citizens is by addressing problems with their nurture. Underfed, unloved, and uneducated human beings have diminished potential to lead productive lives. But as we have seen, nurture, while greatly influential, has its limits, which reveal themselves most dramatically in cases of profound genetic disadvantage. Even with the most perfectly devised nutrition and schooling, boys with severe fragile X disease[1] will still never be able to take care of themselves. Nor will all the extra tutoring in the world ever grant naturally slow learners a chance to get to the head of the class. If, therefore, we are serious about improving education, we cannot in good conscience ultimately limit ourselves to seeking remedies in nurture. My suspicion, however, is that education policies are too often set by politicians to whom the glib slogan "leave no child behind" appeals precisely because it is so completely unobjectionable. But children *will* get left behind if we continue to insist that each one has the same potential for learning.

The guidance of genetic information

We do not as yet understand why some children learn faster than others, and I don't know when we will. But if we consider how many commonplace biological insights, unimaginable fifty years ago, have been made possible through the genetic revolution, the question becomes pointless. The issue rather is this: Are we prepared to embrace the undeniably vast potential of genetics to improve the human condition, individually and collectively? Most immediate, would we want the guidance of genetic information to design learning best suited to our children's individual needs? Would we in time want a pill that would allow fragile X boys to go to school with other children, or one that would allow naturally slow learners to keep pace in class with naturally fast ones? And what about the even more distant prospect of viable germ-line therapy? Having identified the relevant genes, would we want to exercise a future power to transform slow learners into fast ones before they are even born? We are not dealing in science fiction here, we can already give mice better memories. Is there a reason why our goal shouldn't be to do the same for humans?

One wonders what our visceral response to such possibilities might be had human history never known the dark passage of the eugenics movement. Would we still shudder at the term "genetic enhancement"? The reality is that the idea of improving on the genes that nature has given us alarms people. When discussing our genes, we seem ready to commit what philosophers call the "naturalistic fallacy," assuming that the way nature intended it is best. By centrally heating our homes and taking antibiotics when we have an infection, we carefully steer clear of the fallacy in our daily lives, but mentions of genetic improvement have us rushing to run the "nature knows best" flag up the mast. For this reason, I think that the acceptance of genetic enhancement will most likely come about through efforts to prevent disease.

Human resistance to AIDS

Germ-line gene therapy has the potential for making humans resistant to the ravages of HIV. The recombinant DNA procedures that have let plant molecular geneticists breed potatoes resistant to potato viruses could equally well make humans resistant to AIDS. But should this be pursued? There are those who would argue that rather than altering people's genes,

1. a severe combined immunodeficiency disease in male children, which is a genetic disorder that severely limits the immune system

we should concentrate our efforts on treating those we can and impressing upon everyone else the dangers of promiscuous sex. But I find such a moralistic response to be profoundly *immoral*. Education has proven a powerful but hopelessly insufficient weapon in our war. As I write, we are entering the third decade of the worldwide AIDS crisis: our best scientific minds have been bamboozled by the virus's remarkable capacity for eluding attempts to control it. And while the spread of the disease has been slowed for the moment in the developed world, huge swaths of the planet tick away as demographic time bombs. I am filled with dread for the future of those regions, populated largely by people who are neither wealthy nor educated enough to mount an effective response. We may wishfully expect that powerful antiviral drugs or effective HIV vaccines will be produced economically enough for them to be available to everyone everywhere. But given our record in developing therapies to date, the odds against such dramatic progress occurring are high. And yet those who propose to use germ-line gene modifications to fight AIDS may, sadly, need to wait until such conventional hopes turn to despair—and global catastrophe before being given clearance to proceed.

> *Germ-line gene therapy has the potential for making humans resistant to the ravages of HIV.*

All over the world government regulations now forbid scientists from adding DNA to human germ cells. Support for these prohibitions comes from a variety of constituencies. Religious groups—who believe that to tamper with the human germ line is in effect to play God—account for much of the strong knee-jerk opposition among the general public. For their part, secular critics, as we have seen, fear a nightmarish social transformation such as that suggested in *Gattaca*—with natural human inequalities grotesquely amplified and any vestige of an egalitarian society erased. But though this premise makes for a good script, to me it seems no less fanciful than the notion that genetics will pave the way to utopia.

But even if we allow hypothetically that gene enhancement *could*—like any powerful technology—be applied to nefarious social ends, that only strengthens the case for our developing it. Considering the near impossibility of repressing technological progress, and the fact that much of what is now prohibited is well on its way to becoming practicable, do we dare restrain our own research community and risk allowing some culture that does not share our values to gain the upper hand? From the time the first of our ancestors fashioned a stick into a spear, the outcomes of conflicts throughout history have been dictated by technology. Hitler, we mustn't forget, was desperately pressing the physicists of the Third Reich to develop nuclear weapons. Perhaps one day, the struggle against a latter-day Hitler will hinge on our mastery of genetic technologies.

One truly rational argument

I see only one truly rational argument for delay in the advance of human genetic enhancement. Most scientists share this uncertainty; can germline gene therapy ever be carried out safely? The case of Jesse Gelsinger has cast a long shadow on gene therapy in general. It's worth pointing out, though, that contrary to appearances, germ-line therapy should in principle be easier to accomplish safely than somatic cell therapy. In the latter case, we are introducing genes into billions of cells, and there is always a chance, as in the recent SCID case in France, that a crucial gene or genes will be damaged in one of those cells, resulting in the nightmarish side effect of cancer. With germ-line gene therapy, in contrast, we are inserting DNA into a single cell, and the whole process can accordingly be much more tightly monitored. But the stakes are even higher in germ-line therapy, a failed germ-line experiment would be an unthinkable catastrophe—a human being born flawed, perhaps unimaginably so, owing to our manipulation of his or her genes. The consequences would be tragic. Not only would the affected family suffer, but all of humankind would lose because science would be set back.

Leaving aside the uncertainties of [germ-line] gene therapy, I find the lag in embracing even the most unambiguous benefits to be utterly unconscionable.

When gene therapy experiments in mice run aground, no career is aborted, no funding withdrawn. But should gene improvement protocols ever lead to children with diminished rather than improved potential for life, the quest to harness the power of DNA would surely be delayed for years. We should attempt human experimentation only after we have perfected methods to introduce functional genes into our close primate relatives. But even when monkeys and chimpanzees (an even closer match) can be safely gene enhanced, the start of human experimentation will require resolute courage: the promise of enormous benefit won't be fulfilled except through experiments that will ultimately put lives at some risk. As it is, conventional medical procedures, especially new ones, require similar courage: brain surgery too may go awry, and yet patients will undergo it if its potential positives outweigh the dangers.

Serious consideration for germ-line gene therapy

My view is that, despite the risk, we should give serious consideration to germ-line gene therapy. I only hope that the many biologists who share my opinion will stand tall in the debates to come and not be intimidated by the inevitable criticism. Some of us already know the pain of being tarred with the brush once reserved for eugenicists. But that is ultimately a small price to pay to redress genetic injustice. If such work be called eugenics, then I am a eugenicist.

Over my career since the discovery of the double helix, my awe at the majesty of what evolution has installed in our every cell has been rivaled only by anguish at the cruel arbitrariness of genetic disadvantage and defect, particularly as it blights the lives of children. In the past it was the remit of natural selection—a process that is at once marvelously efficient and woefully brutal—to eliminate those deleterious genetic mutations. Today, natural selection still often holds sway: a child born with Tay-Sachs who dies within a few years is—from a dispassionate biological perspective—a victim of selection against the Tay-Sachs mutation. But now, having identified many of those mutations that have caused so much misery over the years, it is in our power to sidestep natural selection. Surely, given some form of preemptive diagnosis, anyone would think twice before choosing to bring a child with Tay-Sachs into the world. The baby faces the prospect of three or four long years of suffering before death comes as a merciful release. And so if there is a paramount ethical issue attending the vast new genetic knowledge created by the Human Genome Project, in my view it is the slow pace at which what we now know is being deployed to diminish human suffering. Leaving aside the uncertainties of gene therapy, I find the lag in embracing even the most unambiguous benefits to be utterly unconscionable. That in our medically advanced society almost no women are screened for the fragile X mutation a full decade after its discovery can attest only to ignorance or intransigence. Any women reading these words should realize that one of the important things she can do as a potential or actual parent is to gather information on the genetic dangers facing her unborn children—by looking for deleterious genes in her family line and her partner's, or, directly, in the embryo of a child she has conceived. And let no one suggest that a woman is not entitled to this knowledge. Access to it is her right, as it is her right to act upon it. She is the one who will bear the immediate consequences.

7

Germ Line Gene Therapy Is Dangerous to Human Health

Paul R. Billings, Ruth Hubbard, and Stuart A. Newman

Paul R. Billings is a clinical genetics expert and cofounder of GeneSage, a company that provides genetic services to consumers and health professionals. Ruth Hubbard is professor emerita of biology at Harvard University and author of Exploding the Gene Myth. *Stuart A. Newman is a professor of cell biology and anatomy at New York Medical College and coeditor of* Beyond the Gene in Developmental and Evolutionary Biology.

In germ line gene therapy, human germ cells would be modified, resulting in inheritable changes in DNA. This has not yet been achieved, but scientific research suggests that the obstacles in conducting the procedure may someday be overcome. However, germ line manipulation must be vigorously opposed for several reasons. It may lead to unintended and undetected changes in gene function that may adversely alter future generations. Also, there is no clinical need for genetic intervention—prenatal diagnosis and other reproductive methods can help couples prevent the transmission of faulty genes. Moreover, germ line gene therapy may be used as an end to eugenics, the improvement of the human race through controlling heredity. To protect future generations from the unforeseen consequences of genetic intervention, germ line gene therapy should not be permitted.

Human germline gene modification has been foreseen but not yet accomplished. It can be defined as the genetic manipulation of human germ cells, or of a conceptus, resulting in inherited changes in DNA. With the development of advanced in-vitro fertilisation (IVF) methods, preimplantation DNA analysis, improved techniques for gene transfer, insertion, or conversion, and of embryo implantation procedures, the technical barriers to such an intervention seem easily surmountable. Unintended changes in DNA may occur when gametes are manipulated or stored. Inadvertent germline mutations, therefore, may have already occurred as a result of reproductive technologies in current use, such as artificial insemination and

IVF. There are unpublished reports that researchers in the USA have already carried out a manipulation involving the exchange of a mitrochondrial genome in an IVF protocol. If true, this human experimentation involving intentional hereditary changes was probably conducted without federal oversight of safety, since there are no discussions of this protocol in the available public record.

Significant burdens

[Researchers T. Tsukui, Y. Kanegae, I. Saito, and Y. Toyoda] used viral vectors in somatic gene therapy protocols to infect mouse eggs in vitro, leading to germline transmission of a transgene in the progeny. Although removal of the zona pellucida[1] is a prerequisite for infection of the eggs in vitro, the early oocytes of postnatal ovaries also lack zonas. These experiments thus raise the possibility that modification of gametes may occur in vivo, and constitute germline hazard in the 200 or more somatic gene therapy protocols [in use in 1999]. Any such alterations would be difficult to detect. Intentional or inadvertent germline modifications may pose significant burdens. Although there are restrictions on experimentation that might result in human modifications, and opposition to its implementation has been voiced, some leading scientists and other commentators have begun to advocate the development and application both of techniques that may increase the risk of inadvertent alteration of the germline, and of methods that would alter it deliberately.

Intentional or inadvertent germline modifications may pose significant burdens.

[Geneticist] W. French Anderson and his colleagues have developed an experimental protocol for the treatment of adenosine deaminase deficiency[2] during fetal development; although their therapeutic intent is directed towards somatic cells, they acknowledge that the technique may modify germ cells as well. They have submitted this proposal to the National Institutes of Health (NIH) for review (panel). By introducing a genetic construct in utero, which knowingly allows for the alteration of germinal tissue, their attempt at a potentially transmissible correction could be used to erode opposition to germline genetic manipulation since germline modification would be achieved, though unintentionally.

Opposition to germline modification is based on several lines of reasoning. First, as we have already suggested, germline DNA modifications may affect gene function in ways that are not immediately apparent, so their occurrence may not be recognised for a generation or more—for example, germline introduction in mice of an improperly regulated normal gene resulted in progeny with unaffected development but high tumour incidence during adult life. Furthermore, interactions among genes and their products are highly integrated, have been refined over evolutionary

---

1. transparent layer or envelope of a mammalian ovum   2. a rare genetic disorder that moderately or completely limits the immune response

time scales, and often serve to stabilise developmental pathways and phys-iological homoeostasis. Through experimental error, unanticipated allelic interactions, or poorly understood regulatory mechanisms such as im-printing, there is a risk that germline genetic manipulation will alter sen-sitive biological equilibria. Disruption of these interactive systems is likely to have complex and uncertain biological effects, including some that ap-pear only during the development or functioning of specific cells or tis-sues. Many of these effects could be undesirable.

No clinical need

Second, this sort of intervention is not needed. With available methods of prenatal diagnosis, virtually all interested couples can choose not to transmit specific identifiable genes. Other reproductive options (artificial insemination, egg donation) and adoption are available to those not able or willing to use prenatal or preimplantation selection methods. An ex-ception might be when, rarely, two individuals have the same recessively inherited disorder. If such couples chose to reproduce, it could be argued that they would "need" germline or very early genetic interventions since all their progeny might inherit a disease-associated genotype. Yet, even these children may differ genotypically and phenotypically from their parents, and the development of a new mode of treatment for this un-usual occurrence does not seem justifiable. Although available alternative procedures are invasive, germline modifications would also require simi-lar interventions, since they would probably involve IVF. Moreover, the associated risks with existing procedures are not as serious as those cre-ated by introducing a hereditary genetic "error" into a family. People who oppose prenatal diagnosis on philosophical or religious grounds would be unlikely to want to take part in germline modification if they were aware of its intrinsically experimental nature and of the numbers of human embryos that would have to be expended during the development of the technology. No unmet need balances the risks of germline interventions to mothers, fetuses, and future generations. Moreover, the costs associ-ated with the general development and implementation of germline ma-nipulation would be formidable.

No unmet need balances the risks of germline interventions to mothers, fetuses, and future generations.

If there is no clinical need for germline modifications, the primary rea-son for using this intervention would be human enhancement. Apart from the uncertainties about its ultimate outcome, enhancement is a form of eugenics. Though not a recrudescence of overtly coercive, public-health-based eugenics popular earlier this century, germline manipulations repre-sent an individual or familial form. Seemingly private personal decisions and "choices" about medical or non-medical programmes for enhance-ment would, nevertheless, reflect prejudices, socioeconomic and political inequalities, and even current fashion. Though enhancement procedures now in use (eg, cosmetic surgery or orthodontics) also change according to fashion, germline intervention would intentionally subject later generations to modifications undertaken on the basis of existing values and conditions. The chance that "desirable" manipulations might later be viewed as disastrous makes germline enhancement "therapies" unacceptable.

Altering the lives of generations

Human germline interventions would necessarily alter the lives of individuals who are yet to be born. Informed consent by the affected individuals is not possible. Extension of the parental right to consent for minors would be required. Such legal permission to specifically alter the lives of generations of unborn individuals would be unprecedented and unjustified.

If germline manipulation is attempted, there will be mistakes or errors in its application. Neither social acceptance nor the necessary range of protections and care for accidentally damaged individuals can be guaranteed. Unexpected alterations in family relationships will occur, and "wrongful life" disputes could arise. Irrespective of whether such interventions were to take place in research or clinical settings, these issues mean that germline modifications cannot be approved by existing standards for the protection of human beings. No benefits to any future individual would justify abrogating or curtailing these restrictions.

Not appropriate or acceptable

For these biomedical reasons, as well as others based in legal, philosophical, cultural, and spiritual/religious traditions, human germline modifications should be opposed and prohibited. Experimentation that may gradually make human germline modification more feasible is under way; it may require further review. Further study is needed of the safety of somatic gene therapy protocols to ensure that they detect, with adequate sensitivity, germline alterations. Many individuals and groups that monitor developments in human genetics can be expected to mount vigorous opposition to the development of human germline protocols, involving direct action, legal manoeuvres, and organising among interested public groups. Unlike many other countries, including those of the EU [European Union], which have prohibited germline manipulation in principle, restrictions on the procedure in the USA are mainly based on practical considerations . . . and are subject to revision as the state of the science changes. Although debate about human germline modifications should continue and, indeed, be broadened to include representation of a diverse cross-section of viewpoints and backgrounds, such discussion should not be construed as suggesting that such a method would ever be appropriate or acceptable.

8

Germ Line Gene Therapy Should Be a Parental Choice

Gregory E. Pence

Gregory E. Pence is the author of Who's Afraid of Human Cloning? *and a professor in the department of philosophy and the school of medicine at the University of Alabama at Birmingham.*

In germ line gene therapy, reproductive (egg or sperm) cells are intentionally manipulated to alter an individual's traits and the genes passed on to his or her descendants. Although human germ line manipulation has not been achieved, it offers the exciting prospects of preventing genetic disorders and enhancing the mental and physical traits of unborn children. Therefore, when it becomes a feasible, safe procedure, parents should have the choice to use germ line gene therapy to enhance their children's health, physical traits, and mental abilities. It is no different than how people today choose reproductive mates that are the most genetically suitable and give their children environmental and educational enhancements that they can afford. Germ line gene therapy as a reproductive choice should not be banned because of misguided fears of genetic engineering and prejudice.

Almost everything that Americans believe about genetic engineering and cloning of humans is false, due to decades of titillating science fiction, sensationalistic reporting in the media, and unthinking opposition. Hence, most people's thoughts and feelings on these topics need education.

Indeed, I personally would like to ban the phrases "test tube baby," "genetic engineering," and "cloning." For the latter, I would substitute the less emotional phrase, "somatic cell genetic transfer," or SCGT.

To assume that germ cells could be modified in a human embryo and have no more risk of harm to the child than natural conception is to remove the only real, moral objection to such procedures. All the other objections to such procedures are either unjustified or surreptitiously assume the resulting child will be harmed in some way, e.g., psychologically by the prejudiced attitudes of others.

There can be no reasonable objection to parents choosing to remove a gene or cluster of genes, or to modify genes, that cause something normally regarded as bad, such as a disease or handicap. Although some disability advocates insist that there is nothing wrong with being deaf, a dwarf, or having Down's syndrome, no reasonable parent would choose to have a child with such a condition when he or she could have a normal child. Indeed, in my opinion, it might be *immoral* to choose to have such a child if one could otherwise have a normal child.

Most people object to letting parents attempt to enhance a child's genotype through germ-cell modification. Usually the hidden assumption is that it really wouldn't work—that something would go wrong—and that the child would be harmed. That takes back the assumption of this essay.

"Perfect children"

The most-repeated objection is that if society let parents make such choices, they would only want "perfect children." Such an objection assumes that ordinary people can't be trusted in creating children. It also implies that wanting the best possible genetic base for a child is a bad motive.

People have not thought this objection through. Men and women exercise choice in selecting mates and in having children. We are quite comfortable with the fact that most of the present six billion earthlings choose the mate they think is the best possible for them and their children. If exercising choice is so bad, why isn't choice about reproductive mates also a dangerous thing? (If we "allow" such a practice, will people want only "perfect" mates?)

Objection [to germline gene therapy] assumes that ordinary people can't be trusted in creating children.

Obviously, what you want and what you get are not the same. As for gene enhancement, it is likely that, for the next decades, we will only have the knowledge to create one trait, especially when its base requires several genes and multifactorial environmental support. As such, parents will have to choose the kind of direction they want to go and decline other directions, e.g., to their child, literary talent but not football talent.

Here is one argument for allowing children to be produced by somatic cell genetic transfer. At least here, we know the cluster of traits that the ancestor had, and many of them may have been genetically based. We may be more likely to get the desired phenotype by reproducing an existing genotype than my fiddling with germline techniques one trait at a time.

Many other objections to attempting human SCGT are based on possible psychological harm to the resulting child from prejudiced reactions of others or from misplaced expectations of parents. We should not ban a reproductive option because some people are prejudiced or misguided. Education is the correct response to prejudice or incorrect expectations, not federal bans.

I do believe that the first attempt at human SCGT should be regulated, in America by a committee such as the Recombinant Advisory Committee

(RAC) at NIH [National Institutes of Health], because the first case is very important to the acceptance of a new option. Louise Brown, the first baby created by in vitro fertilization, fortunately came out healthy, but problems developed in the Baby M case where a surrogate mother was used (and hence, commercial surrogacy was criminalized in some states).[1] So we must be as certain as possible that the first attempt to create a baby by human SCGT will come out well, both for the sake of the child and for the sake of future attempts.

All of this assumes that reproductive science could know one day that germline interventions or somatic cell genetic transfer would cause no physical harm to the resulting child. That is a big assumption. I welcome the day when it is true.

Allowing parents maximal choice

> If you could do so safely, would you use an artificial chromosome to extend the lifespan of your child?

Someday soon, when the opportunities arise, we will see the wisdom of allowing parents maximal choice about their future children. This is not state-controlled eugenics (which attempted to take away such choices from parents), but its opposite. If a child can be given an extra decade of life by an artificial chromosome, or 50 percent more memory through a therapy in utero, then I personally would feel *obligated* to give my future child such benefits. I believe that my child would be grateful to have been deliberately given such a benefit.

Others might disagree and choose not to do so for their children—a decision I would reject. What I fail to understand is how other people— or the federal government—could think it just to prevent me from benefiting my future children in this way, e.g., by a ban on such enhancements (perhaps from a misplaced concern for equality and social justice). I see no difference between such a ban and a similar ban on parents sending their children to computer camps in the summer: both are intended to better children, both will be done most by people with money, and both are not the business of government.

---

1. In 1988, the biological father and his wife were awarded custody of Baby M, an infant conceived through artificial insemination and brought to term by a surrogate mother. The surrogate mother refused to give up Baby M after she was born.

9

Germ Line Gene Therapy Should Not Be a Parental Choice

Bill McKibben

Bill McKibben is a former staff writer for the New Yorker *and author of several books, including* The End of Nature, Maybe One, *and* Enough: Staying Human in an Engineered Age, *from which the following excerpt is taken.*

The rate at which gene therapy is advancing reveals that it may be possible to manipulate the human germ line (egg or sperm cells) to make future generations smarter, taller, and maybe even happier. Although this is an attractive picture, it does not mean that parents should be given the choice to mentally, physically, or emotionally enhance their children through germ line gene therapy. Allowing so would exacerbate, not aid, the pursuit for a better child because genetic enhancements would inevitably become obsolete as genetic therapies improve and individual competitiveness heightens. Making germ line manipulation a reproductive choice would make the pursuit of a better child endless, where moral decisions are replaced by strategic ones.

B y now, the vision of the would-be genetic engineers should be fairly clear. It is to do to humans what we have already done to salmon and wheat, pine trees and tomatoes. That is, to make them *better* in some way; to delete, modify, or add genes in the developing embryos so that the cells of the resulting person will produce proteins that make them taller and more muscular, or smarter and less aggressive, maybe handsome and possibly straight, perhaps sweet. Even happy. It is, in certain ways, a deeply attractive picture.

Before we decide whether all that adds up to a good idea, there's just one more factual question to be answered: Would we actually do this? We've heard from the salesmen making the case, but would we actually buy? Is there any real need to raise these questions as more than curiosities, or will the schemes simply fade away on their own, ignored by the parents who are their necessary consumers and then forgotten by history?

A staple weapon in the cosmetic arsenal

I grew up in a household where we were very suspicious of dented cans. Dented cans were, according to my mother, a well-established gateway to botulism, and botulism was a bad thing, worse than swimming immediately after lunch. It was one of those bad things measured in extinctions, as in "three tablespoons of botulism toxin could theoretically kill every human on earth." Or something like that.

Germline engineering would not promote your own vanity, but instead be sold as a boon to your child.

So I refused to believe the early reports, a few years back, that socialites had begun injecting dilute strains of the toxin into their brows in an effort to temporarily remove the vertical furrow that appears between one's eyes as one ages. It sounded like a Monty Python routine, some clinic where they daubed your soles with plague germs to combat athlete's foot. But I was wrong to doubt. As the world now knows, Botox has become, in a few short years, a staple weapon in the cosmetic arsenal—so prevalent that, in the words of one writer, "it is now rare in certain social enclaves to see a woman over the age of 35 with the ability to look angry." With their facial muscles essentially paralyzed, actresses are having trouble acting; since the treatment requires periodic booster shots, doctors warn that "you could marry a woman [or a man] with a flawlessly even face and wind up with someone who four months later looks like a Shar-Pei." But never mind—now you can get Botoxed in strip mall storefronts and at cocktail parties. "After a brief discussion of benefits and potential risks, everyone starts drinking," explained one doctor who hosts such soirees. "It really takes the edge off."

Less than pressing purposes

People, in other words, will do fairly far-out things for less than pressing purposes. And more so all the time: public approval of "aesthetic surgery" has grown 50 percent in the United States in the last decade, and there's no automatic reason to think that consumers would balk because it was "genes" involved instead of, say, "toxins." Especially since germline engineering would not promote your own vanity, but instead be sold as a boon to your child. Anyone who has entered a baby supply store in the last few years knows that even the soberest parents can be counted on to spend virtually unlimited sums in pursuit of successful offspring. What if the "Baby Einstein" video series, which immerses "learning enabled" babies in English, Spanish, Japanese, Hebrew, German, Russian, and French, could be bolstered with a little gene-tweaking to improve memory? What if the WombSongs prenatal music system, piping in Brahms to your waiting fetus, could be supplemented with an auditory upgrade? According to the *Wall Street Journal*, upscale parents are increasingly buying $18 bottles of baby shampoo, and massaging their infants with "Bonding Oil," an unguent which allows the youngster "to rejuvenate for another day of exploration and growth [according to writer Liz F. Kay]." One sociologist told the *New York Times* we'd crossed the line from parenting to "product development," and even if that remark is truer in Manhattan than elsewhere, it's not hard to imagine what such attitudes will mean across the affluent world. As early as 1993, a March of Dimes poll found that 43 percent of Americans would engage in genetic engineering "simply to enhance their children's looks or intelligence."

Here's one small example. In the 1980s, two drug companies were awarded patents to market human growth hormone to the few thousand American children suffering from dwarfism. The FDA [Food and Drug Administration] expected the market to be very small, so HGH [human growth hormone] was given "orphan drug status," a series of special market advantages designed to reward the manufacturers for taking on such an unattractive business. But within a few years, HGH had become one of the largest-selling drugs in the country, with half a billion dollars in sales. This was not because there'd been a sharp increase in the number of dwarfs, but because there'd been a sharp increase in the number of parents who wanted to make their slightly short children taller. Before long the drug companies were arguing that the children in the bottom 5 percent of their normal height range were in fact in need of three to five shots a week of HGH. Take eleven-year-old Marco Oriti. At four foot one, he was about four inches shorter than average, and projected to eventually top out at five feet four. This was enough to convince his parents to start him on a six-day-a-week HGH regimen, which will cost them $150,000 over the next four years. "You want to give your child the edge no matter what," said his mother.

Shelling out for designer families

A few of the would-be parents out on the current cutting edge of the reproduction revolution—those who need to obtain sperm or eggs for in vitro fertilization—exhibit similar zeal. Ads started appearing in Ivy League college newspapers a few years ago: couples were willing to pay $50,000 for an egg, provided the donor was at least five feet, ten inches tall, white, and had scored 1400 or better on her SATs. (A few months later, a fashion photographer opened a Web site to auction eggs from top models. He offered no guarantees concerning their board scores, saying only, "This is Darwin's natural selection at its very best—the highest bidder gets youth and beauty.") There is, in other words, a market just waiting for the first clinic with a catalogue of germline modifications, a market that two California artists proved when they opened a small boutique, Gene Genies Worldwide, in a trendy part of Pasadena. Tran T. Kim-Trang and Karl Mihail wanted to get people thinking more deeply about these emerging technologies, so they outfitted their store with petri dishes and models of the double helix, and printed up brochures highlighting traits with genetic links: creativity, extroversion, thrill-seeking, criminality. When they opened the doors, they found people ready to shell out for designer families (one man insisted he wanted the survival ability of a cockroach). The "store" was meant to be ironic, but the irony was lost on a culture so deeply consumer that this kind of manipulation seems like the obvious next step. "Generally, people refused to believe this store was an art project," says Tran. And why not? The next store in the mall could easily have been a piercing parlor or a Botox salon. We're ready. And no one's even begun to advertise yet.

A biological arms race

But say *you're* not ready. Say you're perfectly happy with the prospect of a child who shares the unmodified genes of you and your partner. Say you think that manipulating the DNA of your child might be dangerous, or presumptuous, or icky? How long will you be able to hold that line if the procedure begins to spread among your neighbors? Maybe not so long as you think: if germline manipulation actually does begin, it seems likely to set off a kind of biological arms race. "Suppose parents could add 30 points to their child's IQ," asks the economist Lester Thurow, of the Massachusetts Institute of Technology. "Wouldn't you want to do it? And if you don't, your child will be the stupidest in the neighborhood." That's precisely what it might feel like to be the parent facing the choice. Individual competition more or less defines the society we've built, and in that context love can almost be defined as giving your kids what they need to make their way in the world. Deciding not to soup them up . . . well, it could come to seem like child abuse.

If germline engineering ever starts, it will accelerate endlessly . . . as individuals make the calculation that they have no choice but to equip their kids for the world that's being made.

Of course, the problem with arms races is that you never really get anywhere. If everyone's adding 30 IQ points, then having an IQ of 150 won't get you any closer to Stanford than you were at the outset. The very first athlete engineered to use twice as much oxygen as the next guy will be unbeatable in the Tour de France—but in no time he'll merely be the new standard. You'll have to do what he did to be in the race, but your upgrades won't put you ahead, merely back on a level playing field. You might be able to argue that society as a whole was helped, because there was more total brainpower at work, but your kid won't be any closer to the top of the pack. All you'll be able to do is up the odds that she won't be left hopelessly far behind.

An extra ironic twist

In fact, the arms-race problem has an extra ironic twist when it comes to genetic manipulation. The United States and the Soviet Union could, and did, keep adding new weapons to their arsenals over the decades. But with germline manipulation, you get only one shot: the extra chromosome you stick in your kid when he's born is the one he carries throughout his life. So let's say baby Sophie has a state-of-the-art gene job her parents paid for the proteins discovered by, say, 2005 that, on average, yielded 10 extra IQ points. By the time Sophie is five, though, scientists will doubtless have discovered ten more genes linked to intelligence. Now anyone with a platinum card can get 20 IQ points, not to mention a memory boost and a permanent wrinkle-free brow. So by the time Sophie is twenty-five and in the job market, she's already more or less obsolete—the kids coming out of college just plain have better hardware. "For all his billions, [Microsoft chairman] Bill Gates could not have purchased a single genetic enhancement for his son Rory John," writes Gregory Stock. "And you can bet that any enhancements a billion dollars can buy Rory's child in 2030 will seem crude alongside those available for modest sums in 2060." It's not, he adds, "so different from upgraded software. You'll want the new release." The vision of one's child as a nearly useless copy of Windows 95 should make parents fight like hell to make sure we never get started down this path. But the vision gets lost easily in the gushing excitement about "improving" the opportunities for our kids.

If germline genetic engineering ever starts, it will accelerate endlessly and unstoppably into the future, as individuals make the calculation that they have no choice but to equip their kids for the world that's being made. Once the game is under way, in other words, there won't be moral decisions, only strategic ones. If the technology is going to be stopped, it will have to happen now, before it's quite begun. The choice will have to be a political one, that is—a choice we make not as parents but as citizens, not as individuals but as a whole, thinking not only about our own offspring but about everyone. And given the seductions that we've seen—the intuitively and culturally delicious prospect of a *better* child—the arguments against must be not only powerful but also deep. They'll need to resonate on the same intuitive and cultural level. We'll need to feel in our gut the reasons why, this time, we should tell Prometheus thanks, but no thanks.

10

Germ Line Gene Therapy Supports Christian Values

Ted Peters

Ted Peters is a professor of systematic theology at Lutheran Theological Seminary in Berkeley, California. He is also the author of God—the World's Future: Systematic Theology for the World's Future, For the Love of Children: Genetic Technology and the Future of the Family, *and* Playing God? Genetic Determinism and Human Freedom, *from which the following excerpt is taken.*

Germ line gene therapy, which may be possible in the future, refers to the modification of germ (egg or sperm) cells that affect heredity in order to eliminate miscoded genes that cause genetic disorders and diseases. It may also be used for nontherapeutic purposes such as to enhance the mental and physical attributes of future generations. Christian detractors claim that germ line gene therapy is unethical because it usurps the power to create reserved for God. However, germ line gene therapy is ethical within a Christian theology that acknowledges God's creative work as ongoing and humans as the intended cocreator. In addition, the view that humans should not "play God" should not prevent people today from considering genetic technologies, such as germ line gene therapy, as a means of fulfilling the human responsibility to allievate suffering and create a better future.

While wrestling with the interaction of the material and spiritual dimensions of human nature, the midcentury Roman Catholic theologian Karl Rahner described the evolutionary history of the human race in terms of "becoming." Human becoming consists in the self-transcendence of living matter. Nature has a history, and this history develops toward the human experience of freedom in the spirit. But it does not stop there. Nature will progress through and beyond the human stage toward the consummation of the cosmos as a whole, a consummate fulfillment yet to be achieved despite—yet through—the free human spirit. The human race is not merely a spiritual observer of material nature. Nor is human history limited to cultural history. Rather, says Rahner, human history is "also an active alteration of this material world itself." We human beings apply our "technical, planning power of transformation" even to ourselves. As subject we are becoming our own object, becoming our own creator.

Curiously, what Rahner is describing here as human nature is feared by many as usurping divine nature. "Playing God" is the phrase invoked by many to shout "No!" to the attempt by the human race to influence its own evolution.

The acerbic rhetoric that usually employs the phrase "playing God" is aimed at inhibiting if not shutting down certain forms of scientific research and medical therapy. This applies particularly to the field of human genetics and, still more particularly, to the prospect of germline intervention especially for purposes of human enhancement—that is, the insertion of new gene segments of DNA into sperm or eggs before fertilization or into undifferentiated cells of an early embryo that will be passed on to future generations and may become part of the permanent gene pool. Some scientists and religious spokespersons are trying to shut the door to germline intervention and tack up a sign reading "Thou shalt not play God."

The phrase "playing God" is aimed at inhibiting if not shutting down certain forms of scientific research and medical therapy.

Our task here will be to show how the proscription against playing God is being applied to arguments regarding germline alteration, especially the arguments raised by the Council for Responsible Genetics [CRG] in its "Position Paper on Human Germ Line Manipulation." We will see that much of this discussion is thoughtful and wholesome. The ethicists of our day should be congratulated for engaging in pioneering work. I will, however, take the opportunity in this analysis to render a critique of some of the arguments raised against germline intervention. Although I recognize with others that great caution must be taken, I do not believe the dangers call for a lack of vision or a lack of courage. The theological concept of anthropology . . . emphasizes human creativity placed in the service of visionary beneficence, and I think that even germline modification should be considered one possible means oriented toward a beneficent end. I have been arguing that if we understand God's creative activity as giving the world a future, and if we understand the human being as a created cocreator, then ethics begins with envisioning a better future. This suggests we should at minimum keep the door open to improving the human genetic lot and, in an extremely modest way, influencing our evolutionary future. The derisive use of the phrase, "playing God," should not deter us from shouldering our responsibility for the future. To seek a better future is to "play human" as God intends us to.

Somatic vs. germline; therapy vs. enhancement

These issues come to the forefront of discussion due in large part to the enormous impact of the Human Genome Project on the biological and even the social sciences. Descriptively, we know the stated purposes directing the Human Genome Project as presently conceived. First, its aim is knowledge. The simple goal that drives all pure science is present here, namely, the desire to know. In this case it is the desire to know the sequence of the base pairs and the location of the genes in the human genome. Second, its aim is better human health. The avowed ethical goal is to employ the newly acquired knowledge from research to provide therapy for the many genetically caused diseases that plague the human family. [Biomedical ethicist] John C. Fletcher and [gene therapist] W. French Anderson put it eloquently: "Human gene therapy is a symbol of hope in a vast sea of human suffering due to heredity." As this second health-oriented purpose is pursued, the technology for manipulating genes will be developed, and questions regarding human creativity will arise. How should this creativity be directed?

Virtually no one contests the principle that new genetic knowledge should be used to improve human health and relieve suffering. Yet a serious debate has arisen that distinguishes sharply between therapy for suffering persons who already exist and the health of future persons who do not yet exist. It is the debate between somatic therapy and germline intervention. By somatic therapy we refer to the treatment of a disease in the body cells of a living individual by trying to repair an existing defect. It consists of inserting new segments of DNA into already differentiated cells such as those that we find in the liver, muscles, or blood. Clinical trials are underway to use somatic modification as therapy for people suffering from diabetes, hypertension, and Adenosine Deaminase Deficiency.[1] By germline therapy, however, we refer to intervention into the germ cells that would influence heredity and hopefully improve the quality of life for future generations. Negatively, germline intervention might help to eliminate deleterious genes that dispose us to disease. Positively, though presently well beyond our technical capacity, such intervention should certainly actually enhance human health, intelligence, and strength.

Two issues overlap here and we should sort them out for clarity. One is the issue of somatic intervention versus germline intervention. The other is the issue of therapy versus enhancement. Although somatic treatment is usually identified with therapy and germline treatment with enhancement, there are occasions where somatic treatment enhances, such as injecting growth hormones to enhance height for playing basketball. And germline intervention, at least in its initial stages of development, will aim at preventive medicine. The science of enhancement, if it comes at all, will only come later.

Stopping short of endorsement

Every ethical interpreter I have reviewed agrees that somatic therapy is morally desirable and looks forward to the advances gene research will bring for expanding this important medical work. Yet many who reflect on the ethical implications of the new genetic research stop short of endorsing genetic selection and manipulation for the purposes of improving the human species. The World Council of Churches (WCC) is representative. In a 1982 document, we find

---

1. a rare genetic disorder that moderately or completely limits the immune response

> somatic cell therapy may provide a good; however, other is-
> sues are raised if it also brings about a change in germline
> cells. The introduction of genes into the germline is a per-
> manent alteration. . . . Nonetheless, changes in genes that
> avoid the occurrence of disease are not necessarily made il-
> licit merely because those changes also alter the genetic in-
> heritance of future generations. . . . There is no absolute dis-
> tinction between eliminating "defects" and "improving"
> heredity.

The text elsewhere indicates that the WCC is primarily concerned with our
lack of knowledge regarding the possible consequences of altering the human germline. The problem is this: the present generation lacks sufficient
information regarding the long-term consequences of a decision today
that might turn out to be irreversible tomorrow. Thus, the WCC does not
forbid forever germline therapy or even enhancement. Rather, it cautions
us to wait and see. In a similar fashion, the Methodists "support human
gene therapies that produce changes that cannot be passed on to offspring
(somatic), but believe that they should be limited to the alleviation of suffering caused by disease." The United Church of Christ also approves "altering cells in the human body, if the alteration is not passed to offspring."
On June 8, 1983 fifty-eight religious leaders issued a "Theological Letter
Concerning the Moral Arguments" against germline engineering ad-
dressed to the U.S. Congress. The group action was orchestrated by Jeremy
Rifkin of the Foundation on Economic Trends. One member, James R.
Crumley, presiding bishop of the then Lutheran Church in America spoke
to the press saying, "There are some aspects of genetic therapy [for human
diseases] that I would not want to rule out. . . . My concern is that someone would decide what is the most correct human being and begin to en-
gineer the germline with that goal in mind."

---

*The WCC [World Council of Churches] does not
forbid forever germline therapy or even enhancement.*

A more positive approach is taken by The Catholic Health Association. If we can improve human health through germline intervention,
then it is morally desirable.

> Germline intervention is potentially the only means of
> treating genetic diseases that do their damage early in em-
> bryonic development, for which somatic cell therapy would
> be ineffective. Although still a long way off, developments
> in molecular genetics suggest that this is a goal toward
> which biomedicine could reasonably devote its efforts.

The association with eugenics

Part of the reluctance to embrace germline intervention has to do with its
implicit association with the history of eugenics. The term eugenics
brings to mind the repugnant racial policies of Nazism, and this accounts

for much of today's mistrust of genetic science in Germany and elsewhere. No one expects a repeat of Nazi terror to emerge from genetic engineering; yet some critics fear a subtle form of eugenics may be slipping in the cultural back door. John Harris may be a bit of a maverick, but he welcomes eugenics if it contributes to better human health. He makes the point forcefully: "where gene therapy will effect improvements to human beings or to human nature that provide protections from harm or the protection of life itself in the form of increases in life expectancy . . . then call it what you will, eugenics or not, we ought to be in favor of it."

Religious ethical thinking tends to be conservative in the sense that it seeks to conserve the present pool of genes on the human genome for the indefinite future.

Philosophical and ethical objections to eugenics seem to presuppose not therapy but rather enhancement. The growing power to control the human genetic make-up could foster the emergence of the image of the "perfect child" or a "super strain" of humanity. Some religious leaders worry that the impact of the social value of perfection will begin to oppress all those who fall short. Ethicists at the March 1992 conference on "Genetics, Religion and Ethics" said this:

> Because the Jewish and Christian religious world-view is grounded in the equality and dignity of individual persons, genetic diversity is respected. Any move to eliminate or reduce human diversity in the interest of eugenics or creating a "super strain" of human being will meet with resistance.

In sum, with the possible exception of the Catholic Health Association, religious ethical thinking tends to be conservative in the sense that it seeks to conserve the present pool of genes on the human genome for the indefinite future.

Now the question of playing God begins to take on the form of the *Frankenstein* or *Jurassic Park*[2] fever. The risk of exerting human creativity through germline intervention is that, though we begin with the best of intentions, the result may include negative repercussions that escape our control. Physically, our genetic engineering may disturb the strength-giving qualities of biodiversity that we presume contribute to human health. Due to our inability to see the whole range of interconnected factors, we may inadvertently disturb some sort of existing balance in nature and this disturbance could redound deleteriously. Socially, we could contribute to stigma and discrimination. The very criteria to determine just what counts as a "defective" gene may lead to stigmatizing all those persons who carry that gene. The very proffering of the image of the ideal child or a super strain of humanity may cultivate a sense of inferiority to those who do not measure up. To embark on a large scale program of germline enhancement may create physical and social problems, and then we would blame the human race for its pride, its *hubris*, its stepping

---

2. a 1993 film in which dinosaurs are cloned by bioengineers

beyond its alleged God-defined limits that brings disaster upon itself.

Yet, there may be another way to look at the challenge that confronts us here. The correlate concepts of God as the creator and the human as the created cocreator orient us toward the future, a future that should be better than the past or present. One of the problems with the naturalist argument and the more conservative religious arguments mentioned above is that they implicitly assume the present state of affairs is adequate. These arguments tacitly bless the *status quo*. The problem with the *status quo* is that it is filled with human misery, some of which is genetically caused. It is possible for us to envision a better future, a future in which individuals would not have to suffer the consequences of genes such as those for Cystic Fibrosis, Alzheimer's or Huntington's Disease. That we should be cautious and prudent and recognize the threat of human *hubris*, I fully grant. Yet, our ethical vision cannot acquiesce with present reality; it must press on to a still better future and employ human creativity with its accompanying genetic technology to move us in that direction. . . .

The not-yet future and the ethics of creativity

Would a future-oriented theology of creation and its concomitant understanding of the human being as God's created cocreator be more adequate? It would be more adequate for a number of reasons. First, a future-oriented theology of creation is not stymied by giving priority to existing persons over future persons who do not yet exist. A theology of continuing creation looks forward to the new, to those who are yet to come into existence as part of the moral community to which we belong. Second, such a theology is realistic about the dynamic nature of our situation. Everything changes. There is no standing still. What we do affects and is affected by the future with its array of possibilities. We are condemned to be creative for good or ill. Third, the future is built into this ethical vision. Once we apprehend that God intends a future, our task is to discern as best we can the direction of divine purpose and employ that as an ethical guide. When we invoke the apocalyptic symbol of the New Jerusalem where "crying and pain will be no more" (Revelation 21:4), this will inspire and guide the decisions we make today that will affect our progeny tomorrow.

> *"Genetic engineering opens new possibilities for the future of God's creative work."*

The creative component to a future-oriented ethic denies that the *status quo* defines what is good, denies that the present situation has an automatic moral claim to perpetuity. Take social equality as a relevant case in point. As one can plainly see, social equality does not at present exist, nor has it ever existed in universal form. We daily confront the frustrations of economic inequality and political oppression right along with the more subtle forms of prejudice and discrimination that the CRG rightly opposes. Human equality, then, is something we are striving for, something that does not yet exist but ought to exist. Equality needs to be created, and it will take human creativity under divine guidance to establish it plus vigilance to maintain it when and where it has been achieved. [Theologian] Wolfhart Pannenberg, who has developed an ontology of the future, puts it this way: "The Christian concept of equality does not mean that everyone is to be reduced to an average where every voice is equal to every other, but equality in the Christian sense means that everyone should be raised up through participation in the highest human possibilities. Such equality must always be created; it is not already there." An ethic that seeks to raise us to the "highest human possibilities" cannot accept the *status quo* as normative, but presses on creatively toward a new and better future. Applied to the issue at hand, [theologian] Ronald Cole-Turner makes the bold affirmation: "I argue that genetic engineering opens new possibilities for the future of God's creative work."

The created cocreator

We began . . . with an observation of Karl Rahner regarding evolution and human openness toward the future. Self-transcendence and the possibility for something new belong indelibly to human nature. Human existence is "open and undetermined." That to which we are open is the infinite horizon; we are open to a fulfillment yet to be determined by "the infinite and the ineffable mystery" of God. If we try to draw any axioms that connect this sublime theological vision to an ethic appropriate to genetic engineering, then openness to the future translates into responsibility for the future—even our evolutionary future. Such a theological vision undercuts a conservative or reactionary proscription against intervening in the evolutionary process. Rahner describes the temptation to condemn genetic research and its application as "symptomatic of a cowardly and comfortable conservatism hiding behind misunderstood Christian ideals." The concept of the created cocreator we invoke here is a cautious but creative Christian concept that begins with a vision of openness to God's future and responsibility for the human future.

The health and well-being of future generations not yet born is a matter of ethical concern when viewed within the scope of a theology of creation that emphasizes God's ongoing creative work and that pictures the human being as the created cocreator. A vision of future possibilities, not the present *status quo*, orients and directs ethical activity. When applied to the issue of germline intervention for the purpose of enhancing the quality of human life, the door must be kept open so that we can look through, squint, and focus our eyes to see just what possibilities loom before us. This will include a realistic review of the limits and risks of genetic technology. But realism about technological limits and risks is insufficient warrant for prematurely shutting the door to possibilities for an improved human future. Rather than playing God or taking God's place, seeking to actualize new possibilities means we are being truly human.

11

In Utero Gene Therapy Is Dangerous to Human Health

Stuart A. Newman

Stuart A. Newman is a professor of cell biology and anatomy at New York Medical College and coeditor of Beyond the Gene in Developmental and Evolutionary Biology.

Genetically modifying human embryos or fetuses, or in utero gene therapy, has been proposed to prevent the onset of genetic diseases and enhance the traits of unborn children. Such procedures present serious hazards to human health. Although the main objection to in utero gene therapy is that it can adversely alter the genes a person passes on to his or her children, the embryo or fetus undergoing treatment and the woman carrying it are also placed at significant risk. For example, failing to manipulate or deliver genes correctly can harmfully affect development. In one experiment, a mouse's ear, eye, and nose development was disrupted by in utero gene therapy. Also, a pregnant woman whose embryo or fetus undergoes gene therapy can be infected or harmed by genes delivered to her unborn child. Therefore, the unpredictability and hazards of in utero gene therapy make it too dangerous to consider for human experimentation.

The completion of one of the stated benchmarks of the Human Genome Initiative (HGI)—the attainment of a nearly full set of raw human DNA sequences—is certain to give new impetus to proposals to utilize genetics to refashion human biology. The development during the past quarter century of sophisticated in vitro fertilization methods, pre-implantation DNA analysis, improved techniques for gene transfer, insertion, or conversion, and embryo implantation procedures, have placed such interventions on the agenda of biotechnologically-oriented medicine. Currently, the fevered commercial expectations surrounding the HGI over the past decade, along with hyperbole from portions of the scientific community, have lent new urgency to calls for genetic engineering.

Genetic modification of human embryos or fetuses, referred to here as developmental modification, has been proposed for purposes of both prevention of disease and enhancement of capacity. The hazards of genetic modifications to humans have usually been discussed in terms of somatic (body cell) modification, in which only nonreproductive tissues are affected, and germline (egg or sperm cell) modification, in which changes to an individual's DNA can be passed down to future generations. Indeed, this division has led to the general belief that the only, or main, hazard of developmental modification is the potential of transmission of undesired alterations in the germline. But it is clear that the hazards to both mothers and infants of developmental gene modification are much more extensive.

The hazards

The hazards of germline transmission of DNA modification are no longer speculative; the literature on transgenic animals contains numerous examples. For example, germline introduction of an improperly regulated normal gene into mice resulted in progeny with no obvious effects on development, but enhanced tumor incidence during adult life. Such effects may not be recognized for a generation or more.

It is important to recognize that many of these hazards are not eliminated if there is no germline transmission. The biology of the developing individual will still be profoundly altered by the manipulation on his/her genes at an early stage. Laboratory experience shows that miscalculations in where genes are incorporated into the chromosomes can lead to extensive perturbation of development. The disruption of a normal gene by insertion of foreign DNA in a mouse caused lack of eye development, lack of development of the semicircular canals of the inner ear, and anomalies of the olfactory epithelium, the tissue that mediates the sense of smell.

Not only is the "patient" (embryo or fetus) and its progeny at risk from [gene therapy], but so is the pregnant woman.

Attempts at developmental gene modification will certainly be subject to experimental error, but this is not the only source of potentially unfavorable consequences. Certain genes undergo a process of "imprinting" during development, in which the version of the gene inherited from the father or the mother is blocked from contributing to the individual's biological constitution. This phenomenon is part of a wider group of processes known as "allelic interaction" or "paramutation," in which the expression of one version, or "allele," of a gene is influenced by another allele. These phenomena are poorly understood, but it is clear that they are essential to healthy development. Failure of a certain gene to be correctly imprinted, for example, leads to Beckwith-Wiedemann syndrome, which is characterized by organ overgrowth and several different childhood cancers. Simply inserting a desired gene into the embryo in place of an undesired one does not ensure that allelic interaction will proceed appropriately, and experience with farm animal embryo manipulation suggests that it is readily disrupted and results in malformations.

The developmental process is inherently complex, and there is no coherent, scientifically accepted understanding of its overall coordination. And even if this understanding were available, it is clear that the ramifications of developmental manipulation would be inherently unpredictable. For these reasons attempts to genetically alter developed tissues (somatic modification) and attempts to genetically alter embryos (developmental modification) have profoundly different scientific and ethical implications. The tissues of a developed organism are in some sense modular—if blood, or skin, or a heart, or a liver is diseased or damaged it can be replaced by a substitute without changing the "nature" of the individual. Similarly with gene alteration in a developed individual: in reasonable candidate cases the gene is playing a defined and well-understood role in a particular tissue or organ, and the goal of the modification is to replace or correct the poorly functioning gene in one or a very limited set of tissues. Any protocol that sought, in contrast, to introduce into a patient a gene known to have "pleiotropic" (i.e., affecting several systems) physiological effects (a neurotransmitter molecule that mediates communication between nerve cells, for example) would have a difficult time getting approved. It would be like introducing a drug with drastic side effects, but which could not be withdrawn if the patient reacts badly.

No good rationales

During development the situation is even more complicated. During this period, tissues and organs are taking form and the activity of genes is anything but modular. In the course of development almost any gene can have pleiotropic effects, and not just on physiology, but on the architecture of organs, and the wiring of the nervous system, including the brain. One can argue for the use of radical, untested methods to save existing lives, and such arguments, with appropriate informed consent, may indeed justify somatic gene alteration even when scientific experience is still primitive. In such cases, even the failures can legitimately add to the store of useful knowledge. In contrast, there are no good rationales for using untested "heroic" procedures to alter the course of embryonic development except among those who consider that the risks of producing individuals with experimentally produced morphological or neurological aberrations, or increased risks of cancer, are preferable to the options of abortion, or of bearing the unmodified child. . . .

In protocols that attempt somatic "gene therapy" for life-threatening illnesses, saving the life of the individual patient is a value that must be balanced against developmental risks, including those to the germline of that individual, and indeed, such considerations also pertain to chemotherapy in cancer patients, by which mutations may be introduced into the germline. With respect to deliberate developmental modifications, the story is quite different. Not only is the "patient" (embryo or fetus) and its progeny at risk from the procedure, but so is the pregnant woman. If the genes are introduced in utero, such genes can also infect the woman's tissues, including her own germline, and entail other risks to herself, such as cancer. Clearly she is not in a position to give informed consent on behalf of herself or the developing embryo for a procedure that has not yet been tested in humans. In addition, the procedure promises no direct benefits to her health (the usual justification for experimentation on humans). However, she will inevitably be under pressure to assume such risks for the sake of her baby.

Even if the procedure is to be done in vitro rather than in utero, the basis for informed consent remains problematic. There is no existing person whose life is in jeopardy, but rather an embryo in a petri dish that the egg or sperm donor (or whoever else may gain the right to its disposition) would like to modify genetically. No truly informed consent on the part of the potential parents is possible, because no reliable information about the consequences would be available.

Furthermore, no amount of data from laboratory animals will make the first human trials anything but experimental. Under such circumstances, where the life of an existing person is not at issue, and the procedure is inherently experimental—threatening to profoundly alter the biology of the developing individual—contraindication on the basis of safety or unpredictability of outcome (which may be counterbalanced when a life is at stake) becomes an ethical contraindication as well.

Organizations to Contact

The editors have compiled the following list of organizations concerned with the issues debated in this book. The descriptions are derived from materials provided by the organizations. All have publications or information available for interested readers. The list was compiled on the date of publication of the present volume; names, addresses, and phone numbers may change. Be aware that many organizations take several weeks or longer to respond to inquiries, so allow as much time as possible.

American Society of Gene Therapy (ASGT)
611 E. Wells St., Milwaukee, WI 53202
(414) 278-1341 • fax: (414) 276-3349
e-mail: info@asgt.org • website: www.asgt.org

Established in 1996, the ASGT is the largest medical professional organization representing researchers and scientists dedicated to discovering new gene therapies. It is committed to promoting and fostering the exchange and dissemination of information about gene therapy. It publishes a journal, *Molecular Therapy*, and also holds an annual meeting each year featuring scientific symposia, workshops, oral abstract presentations, exhibits, and poster sessions.

Biotechnology Industry Organization (BIO)
1225 Eye St. NW, Suite 400, Washington, DC 20005
(202) 962-9200
website: www.bio.org

BIO represents biotechnology companies, academic institutions, state biotechnology centers, and related organizations that support the use of biotechnology. It works to educate the public about biotechnology and respond to concerns about the safety of genetic engineering and other technologies. BIO publishes the magazine *Your World, Our World.* An introductory guide to biotechnology is available on its website.

Center for Bioethics and Human Dignity (CBHD)
2065 Half Day Rd., Bannockburn, IL 60015
(847) 317-8180 • fax: (847) 317-8101
e-mail: infor@cbhd.org • website: www.cbhd.org

The CBHD is an international education center whose purpose is to bring Christian perspectives to bear on contemporary bioethical challenges facing society. Its publications address genetic technologies as well as topics such as euthanasia and abortion. It publishes the newsletter *Dignity* and the book *Genetic Ethics: Do the Ends Justify the Genes?*

Center for Bioethics at the University of Pennsylvania
3401 Market St., Suite 320, Philadelphia, PA 19104-3308
(215) 898-7136 • fax: (215) 573-3036
website: www.bioethics.upenn.edu

The University of Pennsylvania's Center for Bioethics is the largest center of its kind in the world. It engages in research and publishes articles about many areas of bioethics, including gene therapy and genetic engineering. *PennBioethics* is its quarterly newsletter.

Council for Responsible Genetics (CRG)
5 Upland Rd., Suite 3, Cambridge, MA 02140
(617) 868-0870 • fax: (617) 491-5344
e-mail: info@gene-watch.org • website: www.gene-watch.org

The CRG is a national nonprofit organization of scientists, public health advocates, and others who promote a comprehensive public interest agenda for biotechnology. Its members work to raise public awareness about emerging genetic technologies.

The Hastings Center
Route 9D, 21 Malcolm Gordon Rd., Garrison, NY 10524-5555
(914) 424-4040 • fax: (914) 424-4545
e-mail: mail@thehastingscenter.org • website: www.thehastingscenter.org

The Hastings Center is an independent research institute that explores the medical, ethical, and social ramifications of biomedical advances. The center publishes books, papers, and the bimonthly *Hastings Center Report*.

Institute of Science in Society (ISIS)
PO Box 32097, London, England NW1 OXR
44 20 8643 0681
e-mail: sam@i-sis.org.uk • website: www.i-sis.org.uk

ISIS is a London-based nonprofit organization that promotes both critical public understanding of science and engaging scientists and the public in open debate and discussion. Its publications include *Living with the Fluid Genome* and the journal *Science in Society*. Its website includes many articles on genetics in medicine.

International Forum for Genetic Engineering (IfGene)
c/o Dr. Barry Lia
9314 40th Ave. NE, Seattle, WA 98115-3715
e-mail: barrylia@juno.com • website: www.anth.org/ifgene

IfGene is an organization that explores the diverse views of genetic engineering and the moral and spiritual implications of biotechnology. The forum's website includes many articles on the ethics of genetic engineering, and IfGene's student help desk aids students on assignments, projects, and debates on genetic engineering or biotechnology.

National Institutes of Health (NIH)
National Human Genome Research Institute (NHGRI)
Communications and Public Liaison Branch
Building 31, Room 4B09, 31 Center Dr., MSC 2152
9000 Rockville Pike, Bethesda, MD 20892-2152
website: www.nhgri.nih.gov

The NIH is the federal government's primary agency for the support of biomedical research. As a division of the NIH, the NHGRI's mission is to head the Human Genome Project, the federally funded effort to map all human genes. Information about the Human Genome Project is available at the NHGRI website.

President's Council on Bioethics
1801 Pennsylvania Ave. NW, Suite 600, Washington, DC 20006
e-mail: info@bioethics.gov • website: http://bioethics.gov

The President's Council on Bioethics was formed to advise the president on bioethical issues that may emerge as a consequence of advances in biomedical science and technology. Among the council's other functions are to undertake fundamental inquiry into the human and moral significance of developments in biomedical and behavioral science and technology, to explore specific ethical and policy questions related to these developments, and to provide a forum for a national discussion of bioethical issues. Its publications include *Beyond Therapy: Biotechnology and the Pursuit of Happiness* and *Human Cloning and Human Dignity: An Ethical Query.*

Websites

The following websites contain information that may be useful to students interested in learning more about gene therapy.

Access Excellence
www.accessexcellence.org

Designed for teachers and students, this site offers an overview of gene therapy and links to related topics such as the inheritance of genetic disorders, the Human Genome Project, and biotechnology.

U.S. National Library of Medicine and National Institutes of Health
www.nlm.nih.gov

Sponsored by the National Library of Medicine and the National Institutes of Health, this site offers up-to-date developments in gene therapy clinical trials and research as well general information on related topics such as genetics and genetic testing.

W. French Anderson's Gene Therapy
www.frenchanderson.org

This site contains links to articles exploring the history, science, and ethics of gene therapy. It is the home page of W. French Anderson, the so-called father of gene therapy, and is sponsored by the University of Southern California (USC), where Anderson is a biochemistry and pediatrics professor and director of USC's gene therapy laboratories.

Bibliography

Books

Joseph Alper et al.	*The Double-Edged Helix: Social Implications of Genetics in a Diverse Society.* Baltimore: Johns Hopkins University Press, 2002.
Gavin Brooks	*Gene Therapy: The Use of DNA as a Drug.* London: Pharmaceutical, 2002.
Allen Buchanan et al.	*From Chance to Choice: Genetics and Justice.* New York: Cambridge University Press, 2000.
Audrey R. Chapman and Mark S. Frankel, eds.	*Designing Our Descendants: The Promises and Perils of Genetic Modifications.* Baltimore: Johns Hopkins University Press, 2003.
John Hyde Evans	*Playing God? Human Genetic Engineering and the Rationalization of Public Bioethical Debate.* Chicago: University of Chicago Press, 2002.
Theodore Friedman, ed.	*The Development of Human Gene Therapy:* Cold Spring Harbor, NY: Cold Spring Harbor Laboratory, 1999.
Francis Fukayama	*Our Posthuman Future: Consequences of the Biotechnology Revolution.* New York: Picador, 2003.
Nicholas R. Lemoine and Richard G. Vile, eds.	*Understanding Gene Therapy.* Oxford, UK: BIOS Scientific, 2000.
Glenn McGee	*The Perfect Baby: Parenthood in the New World of Cloning and Genetics.* 2nd ed. Lanham, MD: Rowman & Littlefield, 2000.
Bill McKibben	*Enough: Staying Human in an Engineering Age.* New York: Henry Holt, 2003.
Anders Nordgren	*Responsible Genetics: The Moral Responsibility of Geneticists for the Consequences of Human Genetics Research.* Boston: Kluwer Academic, 2001.
Ted Peters	*Playing God? Genetic Determinism and Human Freedom.* 2nd ed. New York: Routledge, 2003.
Alison Pilnick	*Genetics and Society: An Introduction.* Philadelphia: Open University, 2002.
David B. Resnik, Holly B. Steinkraus, and Pamela J. Langer	*Human Germline Gene Therapy: Scientific, Moral, and Political Issues.* Austin: R.G. Landes, 1999.
Matt Ridley	*Genome: The Autobiography of a Species in 23 Chapters.* New York: HarperCollins, 1999.

G.M. Rubyani and S. Yla-Herttuala, eds.	*Human Gene Therapy: Current Opportunities and Future Trends.* Secaucus, NJ: Springer Verlag, 2003.
Gregory Stock	*Redesigning Humans: Our Inevitable Genetic Future.* Boston: Houghton Mifflin, 2002.
Gregory Stock and John Campbell, eds.	*Engineering the Human Germline: An Exploration of the Science and Ethics of Altering the Genes We Pass On to Our Children.* New York: Oxford University Press, 2000.
James D. Watson and Andrew Berry	*DNA: The Secret of Life.* New York: Knopf, 2003.
Lisa Yount, ed.	*Current Controversies: Genetic Engineering.* San Diego, CA: Greenhaven, 2002.

Periodicals

Ken Adelman	"Changing Who We Are," *Washingtonian*, August 2000.
W. French Anderson	"The Best of Times, the Worst of Times," *Science*, April 28, 2000.
Nell Boyce	"The Cost of a Cure," *U.S. News & World Report*, January 27, 2003.
Penni Crabtree	"Success May Still Lurk in the Genes," *San Diego Union-Tribune*, October 3, 2003.
David A. Dean and R. Allen Perkin	"Gene Therapy: If at First You Don't Succeed . . . ," *American Family Physician*, May 1, 2001.
Tim Friend	"Risky Gene Therapy Gets Personal," *USA Today*, October 8, 2001.
Eithan Gulan	"Gene Therapy Has Vast Potential Despite Setbacks; Push On with Human Trials," *Los Angeles Times*, February 18, 2003.
Irish Times	"Benefits of Some Gene Therapy May Not Outweigh Its Dangers," July 19, 2002.
Eric T. Juengst	"What Next for Gene Therapy? Gene Transfer Often Has Multiple and Unpredictable Effects on Cells," *British Medical Journal*, June 28, 2003.
Leonard R. Kass	"The Age of Genetic Technology Arrives," *American Spectator*, November/December 2002.
Gina Kolata	"In a First, Gene Therapy Saves Lives of Infants," *New York Times*, April 28, 2000.
Stephen Leahy	"Biotech Hope and Hype," *Maclean's*, September 30, 2002.
Joanna Marchant	"Generation Game," *New Scientist*, December 2, 2000.
Marilynn Marchione	"Finding the Right Gene Therapy; Hope for Sound Treatment for Those in Dire Straits Emerging," *Milwaukee Journal Sentinel*, April 28, 2003.
Bruce Murray	"Building Better People: The Truths and Myths of Gene Therapy," *Quills*, June 2001.

Philip Noguchi   "The Risks and Benefits of Gene Therapy," *New England Journal of Medicine*, January 16, 2003.

Sophie Petit-Zeman   "Why We Gambled on Gene Therapy," *Guardian*, August 26, 2003.

Robert Reford   "The Hard Cell: Publicity Surrounding Developments in Gene Therapy May Be Raising Hopes Too Soon," *Nursing Standard*, November 13, 2002.

Judy Seigel-Itzkovitch   "Polishing Up Your Genetic Inheritance," *Jerusalem Post*, May 28, 2000.

Sheryl Gay Stolberg   "The Biotech Death of Jesse Gelsinger," *New York Times*, November 28, 1999.

Jamie Talan   "A Gene Therapy for the Brain; Technique Aims to Quiet Parkinson's," *Newsday*, August 19, 2003.

Larry Thompson   "Harsh Lessons, High Hopes," *FDA Consumer Magazine*, September/October 2000.

Rick Weiss   "Dream Unmet 50 Years After DNA Milestone; Gene Therapy Debacle Casts Pall on Field," *Washington Post*, February 28, 2003.

Jeff Wheelright   "Body, Cure Thyself," *Discover*, March 2002.

Jay Yang and   "Gene Therapy for Pain," *American Scientist*, March 2001.
Christopher L. Wu

L.S. Young and   "The Promises and Potential Hazards of Adenovirus
V. Mautner   Gene Therapy," *Gut*, May 2001.

Index